Agriculture and Rural Development in India
A Case Study on the Dignity of Labour

About the Author

Madan Lal Dewan (Ph.D., Cornell), a well-known Soil Scientist, started work on soil conservation in Damodar Valley Corporation during 1949–52. He had a distinguished career with Food and Agriculture Organisation for over three decades (1953–83). He worked as Chief of the FAO Regional Bureau for Asia and the Pacific. He had FAO assignments in Iran and Bulgaria and was at FAO Headquarters in Rome for 14 years.

Dr. Dewan was visiting Professor and co-ordinator of Himalayan Eco-System Programme with Centre for Policy Research, New Delhi. He was a member of the National Land Use and Conservation Board of Government of India during 1984–87. He is founder Chairman and Trustee of Himcon and has established a Centre "Himalaya Sanrakashan Ashram" at Ranichauri, District Tehri Garhwal, Uttarakhand. *Soil of Iran, Soil Conservation of Iran, Agriculture and Rural Development in India, People's Participation and Rejuvenation of Himalayan Environment* and ten other books are amongst his published works, in addition to more than 100 papers on various aspects of soils, soil conservation, land and water development.

Agriculture and Rural Development in India

A Case Study on the Dignity of Labour

M.L. Dewan

CONCEPT PUBLISHING COMPANY, NEW DELHI-110059

Cataloging in Publication Data--DK
Courtesy: D.K. Agencies (P) Ltd. <docinfo@dkagencies.com>

Dewan, M. L. (Madan Lal), 1922-
Agriculture and rural development in India : a case study on the dignity of labour / M.L. Dewan. -- 3rd ed.
p. cm.
Includes bibliographical references.
Includes index.
ISBN 13: 9788180695902
ISBN 10: 8180695905

1. Rural development--India. 2. Rural development--India--Maharashtra. 3. Agriculture--Economic aspects--India. 4. Agriculture--Economic aspects--India--Maharashtra. 5. Agriculture and state--India. I. Title.

DDC 338.109 547 9 22

This publication is presented in my personal capacity and not as an official of the Food and Agriculture Organization of the United Nations.

ISBN : 13-978-81-8069-590-2
ISBN : 10-81-8069-590-5

First Published 1982

Second Edition with a New Introduction 1987

Third Edition 2009

Published and Printed by
Ashok Kumar Mittal
Concept Publishing Company
A/15-16, Commercial Block, Mohan Garden
New Delhi-110059 (India)

Phones : 25351460, 25351794
Fax : 091-11-25357103
Email : publishing@conceptpub.com

Dedicated

to

the Youth of India

who will help make India more glorious and

reach its potential of being a land of plenty

FOREWORD

Dr. M. L. Dewan's one year's service to the Mahatma Phule Krishi Vidyapeeth, Maharashtra, has blazed a new trail. It presents an intelligent, imaginative, comprehensive approach to problems of agricultural development. His findings should spread far and wide. The involvement of the youth is another feature which has a special relevance today.

M.L. Dewan is a dedicated person. His work is known in many countries. Ever since my days as Director-General of FAO, when I came into personal contact with him, I have entertained high respect for him and his work.

2 December 1981

B. R. SEN
Director General, FAO
(*Retd.*)

Introduction to the Second Edition

Exactly five years back, i.e. in February 1982, the first edition of this book was published by the Concept in a record time of 12 days. This was time to appear at the Tenth International Congress of Soil Science which was held first time in India (and any developing country) where a large number of Soil Scientists from all over the world participated. Since then and almost for last one year, the first edition is out of print and there has been two alternatives – one to reprint it in the same form and the other to up-date it adding some additional experiences of these five years. I have decided jointly with Publishers that we will reprint the same as first edition with a new introduction and some appendices which will highlight some of the recent developments. So there will be a very brief change from the first edition because the additional information and knowledge that has been accumulated deserves a separate publication.

Scanning over five years, starting from February 1982, the Soil Science Congress was a great success better than most of foreign guests expected. It also gave a stimulus and inspiration to Soil Scientists and Agricultural Scientists throughout the country. Looking at Appendix VII of the first edition, the readers would recall that I was privileged to be responsible on behalf of Soil Science Society of India for inviting the Congress to hold their Session in India.

I was Chief of Regional Bureau for Asia and Pacific at the FAO HQ, Rome from 1975 to 1982 and from 1st April 1982, I became the Chief of TCDC (Technical Cooperation among Developing Countries), Inter-regional Global and Special Programmes Unit in the FAO HQ, Rome. It was an additional and interesting experience where I could make some contributions which were outgrowth of my previous experiences. I also travelled a lot in 1982-83, around the world, also as the guest of Chinese Government – both myself and my wife and we travelled and experienced their great hospitality. This trip also took us to Brazil, Peru and Mexico.

I started a TCDC News Bulletin from FAO HQ, Rome and brought out several issues before I retired on 15th July 1983. In this period, I also had the privilege to participate in a meeting on behalf of FAO which was held at the UN General Assembly Hall in May 1983 where I addressed the meeting on the subject of TCDC. I had also participated in an African TCDC meeting in West African City of Librevillee, Gabon (West Africa). I recall my introductory address to this meeting where I mentioned that although I am the son of Asia, I am the son-in-law of Africa because my wife was born and brought up in Nairobi, Kenya.

Although 1982-83 was very exciting but greater excitement awaited me on my return after retirement back to my country arriving on 1st October 1983 after about 31 years of service with the FAO.

I have the privilege to be invited as a Member of the National Land Use and Conservation Commission later renamed as National Land Use and Conservation Board which I continue as such.

One of the major activities has been the Himalayan Eco-System Development. Centre for Policy Research (CPR) also hosted me as a Visiting Professor at CPR and to work on the Himalayan Eco-System Development. The first meeting of NGOs with participation of the government departments was held in March 1985 wherein quite a number of NGOs participated and we brought out a publication from CPR entitled "People's Participation in Himalayan Eco-System Development". Since then, some pilot projects, some new NGOs ready for work in the Himalayas and some other activities started. Then the National Land Use and Conservation Board invited me to prepare a report on "The State of Himalayas 1986", I added to it "A Call for Action" and this report has been presented to the Board, Planning Commission and to various other bodies and hope that some increased action for conservation and development of the bleeding Himalayas will steadily and hopefully speedily emerge.

In this period, I had the privilege of starting, together with Mr. C.L. Bhatia, Inspector General of Forests (Retd.) and many other Conservationists, a new organization called PURE (Plantation, Utilisation and Reforestation Enterprise). Many non-governmental organizations have started working in the field of tree plantation, environmental control, conservation and wasteland development, etc. I also have the privilege to work with the SPWD (Society for Promotion of Wasteland Development) as their Technical Adviser and a member and I continue to collaborate with many other non-governmental organisations such as CORT (Consortium on Rural Technology) which also elected me as Vice-Chairman; CSE (Centre for Science and Environment) (I was a member of the Executive Committee for about 2 years); Foundation for Critical Choices for India, Delhi Chapter (I was elected as Chairman) and many other Societies.

I have participated and helped FAO in preparation of projects for the governments in the fields related to conservation and development, land and water development, aerial seeding and now I am leader of the FAO/ICAR/KVK Mission which is expected to give suggestions on the future expansion both in width and depth of the Krishi Vigyan Kendras – Farm Science Centres or popularly called KVKs. The mission is expected to give its report by end February 1987.

One other experience I can share in this introduction is being the Chairman of the Narmada Watershed Committee which was a committee of representatives from Madhya Pradesh, Maharashtra, Gujarat and the Central Government. The experience was very heartening but less than adequate action has emerged. It is to some extent discouraging but being determined to see some conservation policy and programme still accelerates in the Narmada basin, I feel that it is not a given up case.

I have participated in several meetings related to non-conventional energies and addressed them. But the most heartening is the recent development on biogas promotion and more especially biogas from human waste. I was invited to be the FAO/UNDP Training Coordinator of the Fifth International Biogas Training Course at Chengdu, Sichuan, China held in April-May 1986. The report on this training course which I brought out is worth going through and is giving a hope of much increased work in the field of biogas.

The last experience which I like to share with the readers is the experience of actually starting an Integrated Farm in Delhi State, a small six acre one, which combines tree planting, horticulture, vegetable farming, animal husbandry including dairy farming and fish farming and fodder cropping plus flowers and above all a nursery whereby I spread the tree planting by providing seedlings of fuel, fodder and adopted fruit species to the potential users.

This farm has also a clinic which can help the neighbouring villagers in providing medical help. It has given me opportunity to work in the neighbouring villages of Najafgarh block and thus be of service to the villagers.

Many new things await the coming years; a computer system for the whole Himalayan Eco-System Development and a dynamic and integrated programme for saving the Himalayan involving dozens of government and non-government organizations and thousands of young people and planting millions of trees and doing other conservation work. The field is vast but life time is small; ready to continue till the health permits and I am lucky in this respect. Hope that more and more of FAO and UN returnees as well as retirees from the government service and also the non-resident Indians returning to their motherland can make a wonderful contribution to the development of our rural society (and urban society) because we as a Nation are still poor even though we have abundant natural and human resources. Still the hope for such development lies in the youth of India.

This second edition is re-dedicated to the young men and women of the country and more especially of the Himalayan Districts for their determination to safeguard the Himalayas and stop the sacred mountains from bleeding and crumbling.

N-44 Panchshila Park
New Delhi-110017
26 January 1987

M.L. Dewan

INTRODUCTION TO THE THIRD EDITION (2008)

"Agriculture & Rural Development in India : A Case Study on Dignity of Labour" although first edition came in 1982, and second edition in 1987. Twenty years have passed and the publisher still reminded me of bringing out a third edition. I reflected, twenty years since have the findings and the text not become out the date. I went through it again and updated it with the selected work of Agriculture and Rural Development and specially, the dignity of labour, I find that many tasks accomplished, findings and recommendation continue to be relevant even today.

I am not comparing it with the old scripture but still humbly feel that we should bring out this third edition as publisher advises there is some demand and hope that it can be useful to the youth. Hence this Edition is dedicated to the youth of India who will help India more glorious and reach its potential of being a land of plenty. Although recent events and developments reflect some negative factors in Agriculture and rural development. However, some factors and solution still hold such as "Large Scale Tree Planting", "Water Harvesting", "Global Warming Reduction", "Organic Farming and Volunteering" by young and not so young, Agriculture and Rural Development, Science in India will help and is still effective for a great rural population of the country. To the Youth, an appeal, Believe in Dignity of Labour, practices it and work in Rural areas with the farmers and farm women as volunteers, partners and even co-farmers and bring up real and effective agriculture and rural development in the country. I still have confidence in the Rural Population that we must continue to have self-confidence and confidence in Agriculture and Rural Development in India as the Saviour and Hope of India's Sustainable Development and Sustainable Society.

M.L. Dewan

Introduction

Maharashtra, with a population of about fifty-two million, has been striving to be self-sufficient in her food requirements. In the meanwhile, she has been importing annually 200 to 300 crores worth of foodgrains. Obviously, the imports cannot go on for ever and the State must produce her needs. Increased agricultural production and productivity is not merely a function of successfully growing hybrid and high-yielding varieties along with their package of practices. Basically, the soil, water and other physical resources will have to be fully conserved and wisely utilized. Maharashtra's food problem would be nearer to its solution when a balance is struck between the soil-water-plant relationship, especially at the farmers' level.

Dr. M.L. Dewan during his stay of one year with MPKV has striven hard to emphasize the proper use of soil, water, land and manpower resources of the State. In the Report that he has presented, he has interpreted the soil of MPKV's campus at Rahuri, reviewed the land resources of Ahmednagar district and has touched on the resources of Maharashtra State as a whole. He has strongly emphasized the use of organic manures, has laid stress on social and farm forestry and has made suggestions for making the DPAP project a grand success. He has also put up proposals for the establishment and conduct of a Maharashtra Institute of Natural Resources. He has outlined various steps for the development of the Institute. Both by example and precedent, he has advised on utilizing the land and the power of youth of our schools and colleges, including drop-outs.

On behalf of MPKV, I wish to place on record our fullest appreciation of the contribution that Dr. Dewan has made during his brief period of one year's stay. We are indebted to Mrs. Dewan for taking a great deal of interest in the education of the children of the working class and the white-collared parents. By and large, the brief stay of Dr. and Mrs. Dewan has had a very healthy influence on the campus life.

18 April 1975

M.S. Pawar
Vice-Chancellor
Mahatma Phule Krishi Vidyapeeth
Rahuri

Preface

It WAS some time in December 1973, in Sofia, Bulgaria, where I was working as FAO Project Manager at the Poushkarrov Institute of Soil Science as well as FAO Country Representative, that I read in the papers coming from India about the discouraging food situation in my country. There were near famine conditions in some parts. In contrast, in Bulgaria the average wheat yields for the country were touching 3.8 tonnes per hectare and maize yields about 7 tonnes of grain per hectare.

I was disturbed mentally, emotionally and technically as I had been previously several times. This time I thought: "Here is an Indian Soil Scientist, as International Civil Servant, who is assisting a country which is doing much better than his own". The limitations of the individual are known, as also that of an expert away from home. However, I decided to offer my services in an honorary capacity to India, preferably to an Indian Agricultural University where I could try some of my ideas in the field. These ideas, I thought, might open some vistas which may help India even to a small extent, in her vast problems in the field of food and agriculture, and especially in land and water development. I was, and am, convinced that the country's vast natural resources, particularly in the field of land and water development are underutilized, even misutilized, and feel that something can and must be done to improve the situation.

In this process, I wrote to a Vice-Chancellor friend, Dr. M.S. Pawar. We had been in touch with each other since we left Cornell University, USA, where we had been together as post-graduate students from 1946 to 1948. I offered my services for one year in an honorary capacity to work in the field of Agricultural Development in the context of national development, and with special reference to soils and land and water development. His reaction was an immediate invitation to come to Mahatma Phule Krishi Vidyapeeth and help in establishing the Soil Institute of Maharashtra which was under consideration by the State Government. So I decided to go back to India and serve my country.

As it was found on arrival that due to budgetary problems, there was no allotment for the proposed Soil Institute of Maharashtra, a variety of basic tasks, some as agreed and some requested by the Vice-Chancellor, were undertaken, as follows:

1. Dignity of Labour Camps for Students.
2. Soil and Land Resources Evaluation.
3. Land and Water Resources Planning and Development (Special request of the Vice-Chancellor to prepare a Master Plan for Land and Water Development of the MPKV Central Campus, Rahuri).
4. Irrigation, Drainage and Reclamation in the Mula Command Area.
5. Organic, and Rural and Urban Composting in Agricultural Development.
6. Bone-meal for Augmenting Phosphate Supply.
7. Wheat Yield Programming Project.
8. Drought Prone Area Programme Projects.
9. Youth Corps Activities – Participation in Youth Camps, etc.
10. Social and Farm Forestry.
11. Proposals for the establishment of the Maharashtra Institute of Natural Resources for Agriculture which was developed according to the felt needs of the State and piloted for approval by the State Government.
12. Training Programme in Soil and Water Management, Conservation and Development.

The following pages describe some of the tasks undertaken. Nana (Dr. M.S. Pawar), as Vice-Chancellor of a dynamic and upcoming Agricultural University, provided all the facilities at Rahuri where I was stationed. Rahuri is a small place, but then there was the excitement of an action programme, a creative programme with team spirit in collaboration with members of other disciplines.

I took an undertaking that I would make everyday a new day, with new tasks and innovations. I would innovate, give leadership, coordinate and do as much of physical and mental work as possible, I would not get involved in administrative and financial matters for there is no time to be lost.

It is not easy to work in an Indian system. There are greater incentives for those who do not work. The atmosphere in MPKV was different and this is where proper and devoted leadership play such an important role. I was, therefore, happy and felt myself fortunate to be a part of this university where a transformation was taking place. I am not unduly impatient, but I did often feel that things might have moved faster if all had taken an active part. But this may be idealism, because we know that in life there is a minority of "Givers" and a vast majority of "Takers".

I travelled a lot in the country during that year, mainly to learn, but in this process I gave several suggestions and recommendations, most of which were accepted both in Maharashtra and outside, especially in the DVC (Damodar Valley Corporation in Bihar and Bengal) area where I was invited by the Chairman to help revitalize the soil conservation department of DVC (where I had worked in the early fifties).

I attended and addressed several meetings of scientists, of technical personnel, of students and farmers. Some of the talks and papers are reproduced in the Appendices.

Time will be the judge of what has been achieved. The description that follows shows what was attempted. In scientific endeavours and efforts, a scientist does not usually use the name of God. I am, however, convinced that it was God's design for us to come here to serve. How it all happened that by a combination of circumstances we could get one year's leave from FAO to serve India, Maharashtra and MPKV is more than I can explain. I am grateful to Nana Pawar who invited, encouraged and gave us all the support he could, and more, to make this mission a success. He gave this support with affection, kindness and love in abundance and went along with all the ideas I presented. I should also like to express my gratitude to Dr. M.S. Swaminathan, who always spared time to discuss, to exchange views and to give and accept advice, and give all encouragement and support.

I am affectionately grateful to my wife and children. My wife's willingness and participation in the activities of the University and her contribution made our stay more interesting and useful. She put in her best for human resource development, especially with regard to children and women, in spite of her ill health in the second-half of our stay. My son worked in the Dignity of Labour Camp and also prepared projects for youth training and youth corps for national development. I am most grateful to FAO for giving me one year's leave, as requested.

I am also much indebted to a large number of persons who have assisted me in my programme of work. Prof. D.N. Surve, Mr. J.H. Dongale, Mr. Puranik and Mr. T.D. Patil all assisted me technically. My stenographer, Mr. V.N. Kulkarni, worked hard. Mr. Kasar, PA to VC, and Mr. Mistry also helped a lot. I am grateful to them all for their willing and enthusiastic assistance.

In executing each programme indicated in the chapters that follow several people cooperated closely with me. The list is too large to mention each one individually but my grateful thanks are to all of them. In general, I wish to state my gratitude to the University authorities, to the Vice-Chancellor, to the Dean of Agriculture Faculty, the Dean of Lower Agriculture Education, all Associate Deans and Heads of the Departments, and also the Registrar, Comptroller, Deputy Registrar and the Director of Farms and Farm Superintendents.

My thanks also go to the State Government headed by the Chief Minister, the Ministers of Agriculture and Irrigation, the Chief Secretary of Planning and many other officials, especially the Director of the Bureau of Economics and Statistics, Dr. M. Telang, and the Joint Secretary of Planning, Mr. Tungare, who offered special cooperation in steering the proposal of MINRA. With the Departments of Agriculture, Animal Husbandry, Irrigation especially Irrigation Research and Development (Dr. Ratnaparakhi), Forestry, and Education, I cooperated in the relevant fields and I received their assistance.

Above all, the farmers of the villages around the MPKV Campus, such as Digras, Sade, Wanjarwadi and many other villages of the Mula Command Area and villages from other parts of Maharashtra, where we carried out the wheat yield and other development or training programmes, deserve special mention. It was a joy to meet them, talk to them and serve them, and to receive their response and hospitality and kindness.

The report is submitted to MPKV, to the State Government Minister of Agriculture and ICAR, Government of India, and I hope copies will be made available to all interested in the universities, in the State Government Departments and other Government and non-Government bodies interested in land and water development and general agricultural development. This report is a case study of what can be done by team work and is open to a discussion on the philosophy and technology of agricultural development.

It is hoped that a large number of Indian scientists and specially agricultural scientists working abroad, even if they cannot come for a long period, can spend part of a year or more in their country and give the benefit of their knowledge and experience in a spirit of devotion, to see for themselves the problems which exist and help in their solution to the extent possible.

The Report was published by MPKV in a limited number of copies. Although there had been a demand, since, for more copies, a reprinting was not done, partly as it was considered preferable to wait for some five years and see if the ideas presented in the Report would still have validity and usefulness. However, as requests continued to be made, and also encouagred by the kind Foreword prepared by Dr. B.R. Sen, unasked, I decided to have a larger number of copies printed.

In this volume, a Part II has been added: "Increasing National Efficiency in India" as well as some proposed Plans of Operation. Part II is based on a still earlier paper I prepared in 1965. Going through this, I was struck by the fact that what I had written more than fifteen years ago, is still valid – and even more so – for the agricultural, rural and general development of my motherland, India. I feel ever more convinced that India's vast potential of resources is grossly underutilized.

In preparing the Report and its reprint I owe a lot to the confidence of the following five persons :

(1) Dr. B.R. Sen, ex-Director-General of FAO, who has given me his friendship, love and affection;
(2) Dr. M.S. Swaminathan, Member of the Planning Commission, whom I have always admired for his knowledge, his devotion to India and his humility, and whose friendship and readiness to exchange views and experience I have valued greatly;
(3) Dr. M.S. Pawar, who has been a friend and colleague over the last thirty-five years;
(4) Mr. P.S. Mohan, who has helped me in improving the presentation of the text by editing both Part I and Part II.
(5) My wife Darshi, who has been a great source of inspiration and offered me her devoted partnership during my one year's stay at MPKV, Rahuri – a year on which I look back in fond reminiscence.

This document is printed/reprinted in my personal capacity and not in that of an official of the Food and Agriculture Organization (FAO) of the United Nations. I am, however, greatly indebted to FAO for affording me a life-time experience under a variety of conditions, culminating in the present one as Chief of

Regional Bureau for Asia and the Pacific in the Field Programme Development Division where I am privileged to serve countries of Asia including my own, India.

Villa Gitanjali
98, Via Basile
Rome 00128, Italy
January 1982

M.L. Dewan

Contents

Part-II
Plans of Operation for Increasing National Efficiency in India (1965)

List of Plates

Part-I

One Year's Service to an Agricultural University, Maharashtra and India (1974-75)

1 TFL – Trees for Life

Trees for Life (TFL) was founded as a World Organization by Balbir Mathur, with a Trust registered at Allahabad. The Himalayan Operations are directed by Dr. M.L. Dewan, Adviser, TFL, and Founder Chairman, HIMCON (Himalayan Consortium for Himalayan Conservation).

Trees for Life provides management and know-how to people to plant and care for food-bearing trees to fight hunger and malnutrition. Trees for Life is a non-profit organization.

Trees of Life was started with the following objectives:

(i) To provide techniques and materials required to plant, cultivate and harvest food bearing and other trees and to undertake plantation and/or assist others in plantation of such trees.

(ii) To carry out programmes and talks to public places, schools, and service clubs etc. to provide education, training infusion of appropriate technology and awareness amongst people for *plantation of trees to fight malnourishment, hunger and for improvement of ecosystem.*

(iii) To develop a series of clear, concise and candid materials for the purpose of being made available to all such public organizations and societies who are genuinely concerned with the problem of fighting hunger and malnourishment, so as to enable them to convey the message and necessity of planting more food bearing and other trees in the most correct manner.

TFL–HIMCON

Focusing on the fragile eco-system of the Central Himalayas, HIMCON, in conjunction with TFL, and other voluntary and people's organizations is attempting to find alternative development approaches to bring about a qualitative change in the lives of the simple communities through their own active involvement and participation. Thus HIMCON as partner of TFL in the Himalayas enlarging the scope and activities of TFL in the central and western Himalayas, is following the strategy well established for horticulture growth and development of the region.

HIMCON

HIMCON is a non-profit voluntary organization which was initiated by its founding Chairman Dr. M.L. Dewan more than 15 years ago under the banner of TFL-HIMCON. It was later registered in New Delhi as an independent charitable trust called HIMCON initiated by Mrs. Sudarshana Dewan in the year 1999. The Dewan family started its Himalayan networking from their home at Shivranjini later also called as Himalaya Surakshan Ashram at Ranichauri in the Tehri-Garhwal district of Uttarakhand and developed it as its field headquarters for exchange, development and implementation of new ideas towards improving the quality of life of the Himalayan people and its environment.

Our Focus

The HIMCON team is dedicated to the Protection and Conservation of Himalayas and focuses on:

- Improvement of the quality of life of the masses of this region through small initiatives originating from within the society and mostly supported by it.
- Greening the Himalayan terrain.

- Organic farming and horticulture development.
- Propagation of medicinal plants and herbs.
- Water harvesting and conservation.
- Providing safe drinking water to the village communities.
- Improving the nutritional value of the food available to the masses.
- Land slide mitigation.
- Value based education.
- Vocational Training.
- Traditional Health Care and Nature Cure.
- Motivating youth and women to participate in various developmental activities.

MOUNTAIN INDIA PEOPLE'S FORUM – MIPF

Mountains are many things to the people, place of pilgrimage, home of Gods, 'natural resource repositories, bio-diversity treasure groves, source of water, aesthetic, landscape and much more', however, the most important aspect is that they are home to many people of diverse cultures.

United Nations Conference on Environment and Development acknowledged that mountains are often net exporters of globally significant natural resources to the low lands below.

The work of the Mountain India People Forum will be directed towards securing better quality of life for people living in the mountains. It is estimated that 1/10 of homes are in mountains and 25 per cent of landscape is mountainous in addition 2 billion people depend on mountains. Findings from all over the world show that expanding economic pressures are degrading mountain eco-systems, while confronting mountain people with increasing cultural assimilation, poverty and often political disempowerment.

Objectives

The key objectives of MIPF are:

(a) To promote sustainable development of upland and mountain people and their environments.
(b) To promote linkage and networking between stakeholders in people based development of mountains and upland regions.
(c) To promote individual initiative to create awareness regarding poverty, unemployment, poor health, and education and inadequate work for the improvement of the mountain communities.
(d) To undertake, organize and facilitate action programme based on people's need to involve their participation in planning and implementation.
(e) To undertake, organize and facilitate training courses, conferences, seminars, lectures and research in the matters relating to sustainable development of the people of the mountains and upland regions and the preservation and improvement of their natural and cultural heritage.
(f) To undertake and facilitate publication of newsletter, research papers, books, journals and reports, and audio visual material etc. for promotion of the cause and understanding of mountain and upland people.

MOTI – MILLIONS OF TREES INTERNATIONAL

Concept - to Heal the Earth for Healthy Life

We know the whole life of trees is to serve the humanity.

(i) With their leaves, flowers, fruits, branches, roots, shade, fragrance, sap, bark, wood and also their ashes and coal, they exist for the purpose of others.
(ii) A tree that lives for 50 years generates Rs. 10.3 lakhs.
(iii) Recycles Rs. 6.4 lakhs worth of fertility and soil erosion control.

(iv) Creates Rs. 10 lakhs worth of air pollution control.
(v) Rs. 5.3 lakhs worth of food and shelter for birds and animals.
(vi) Besides it provides flowers, fruits and timber.

So when even one tree falls or is felled, the community loses something worth "more than Rs. 32 lakhs."

Nature has always inspired and challenged human life. The trees are an essential component of life and play a critical role in sustaining the natural and human environment. But the human interventions in nature have created unprecedented environmental hazards. Felling of trees has created tremendous eco-imbalance and if it is not taken care of now, humanity and other creatures are condemned to a disastrous future.

The very survival of planet Earth depends on saving trees from destruction.

"Millions of Trees International" is working for a healthy world and healthy humans. It is a non-profit organization working for creating awareness, facilitating tree plantation with the vision to synergize human life with nature for a healthy and happy future.

Mission

1. Restore our natural heritage.
2. Involve the common man specially the young minds to be a part of making the natural environment healthy.

Our Methodology

1. Involvement of Women led Voluntary Agencies for leading such a movement.
2. Identifying spots and organization having sufficient and suitable land for nurseries.
3. Identifying Schools, Colleges and other interested institutions as partners.

4. Involvement of School and College students and to promote a culture for planting and protecting trees.
5. Develop a linkage with organizations such as "Trees for Life" founded by Balbir Mathur who is leading a Global Awareness and Action Programme for Tree Planting and Protection.
6. Develop a link with Corporate Bodies and their Associations such as CII, ASSOCHEM, PHD, FICCI and other national and international associations for promoting greening of their establishments and helping voluntary agencies in such mission.

Activities

1. Organizing "Tree Plantation" Programme mainly in different schools through their management in school campuses in all areas.
2. Organizing 'on the spot' Painting Competition, Quiz Competition and different Games which will create more awareness and interest among the school children.
3. Awarding prizes to the best schools who will plant maximum trees and protect them also.
4. Making Tree Planting a Movement with Women's Groups, NGOs including rural and urban areas that need such tree planting.
5. Organising Tree Plantation in Corporate Land and by Corporate Organization.
6. Rejuvenation and Regreening of Wastelands.
7. Participate in, "OWN/GIFT/DEDICATE A TREE" Project as a part of this Movement.

We intend to accomplish the above objectives through various programmes and projects in different institutions and hence increase the awareness among all specially the young hearts and minds towards such pressing needs of the country, the nation and the world.

Temperate Fruit Planting in Himalayan Region – Help and Cooperation of Dewan Foundation

Introduction

Newly created state of Uttarakhand represents essentially a Himalayan state that extends from the Indo-Gangetic alluvial plains in the south to Higher Himalaya in the north. With 13 administrative districts, the state has two divisions – Garhwal and Kumaun. During the British rule Kumaun division was under direct British rule while only a part of the Garhwal division was under direct British rule. Tehri-Garhwal district of Garhwal sub-division was part of the protectorate princely state of Tehri-Garhwal and the policy of isolating the region from other parts of the country followed by the then rulers of the state has resulted in this region lagging behind in the spread of modern education and others. As a consequence of the legacy of the rulers of the princely state this region has lagged behind in developmental initiatives and is still backward compared to other regions of the state.

The district is drained by Bhagirathi and Bhilangana rivers. Alaknanda forms the southern boundary of the district and meets Bhagirathi at Devprayag, downstream of which the river is called Ganga. Yamuna river demarcates the western boundary of the district with Dehradun.

Rishikesh-Gangotri national highway passes through the district and the district has a considerably good network of link roads.

The district has eight developmental blocks, namely Dhauladhar, Pratapnagar, Bhilangana, Jakhnidhar, Kirtinagar, Deoprayag, Narendranagar, and Chamba.

The Region

Though rugged Himalaya provided basic ingredients for the growth of rich and diverse cultures in its lap and through experience, experimentation and accumulated knowledge there evolved a rich tradition of resource management that was well suited to the ground realities of the terrain. These cultures set forth foundations of a sound system of resource management and developed through their knowledge of genetic craftsmanship a whole range of farm and animal species well suited to the agro-climatic conditions and physiography of the region.

Economy of the region was essentially based upon animal rearing and the forests of the region supported large cattle herds. Agriculture was practiced in river valleys as also in the mountain slopes that were terraced for this purpose. Economic strength of the traditional system is reflected in majestic structures that were supported by its surplus produce and even the common man owned these. The people initiated a well thought of system of water management taking care of both increasing recharge and at the same time avoiding mass movement by increased pore water pressure. The people were aware of the intricacies of the hydrological cycle.

However, with the onslaught of modern western thought, consequent breakdown of the social solidarity, and withdrawal of the state support the precepts of traditional system of resource management were relegated to the background. This is reflected in enhanced frequency of natural disasters and high rates of mass wastage in the region; apart from a number of ill effects in the plains. The surplus producing economy of the region was consequently transformed into the one dependent upon outside remittances, now dubbed as a "Money Order Economy". As a reaction to the changes taking place and in a bid to strengthen their economy and also to improve their quality of life the able-bodied personnel of the region resorted to out-migrate in search of jobs, so as to create surplus to be pumped into the cash starved economy of the region.

Though well-intentioned, this option further destabilized the economy and the region as with out-migration the region faced shortage of farming hands and the lot of the females was highly overburdened; they have to take care of the various farm related activities besides the routine household chores and collection of water, fodder and fuelwood. The fairer sex could hardly manage all these efficiently and they were forced to prioritize the various works that they had to do. The agricultural initiatives in the far-flung fields were abandoned and the land was ill-managed. This led to depleting land fertility, increasing erosion. The reality is that the region faces severe dearth of working hands, brains, money, and youth.

Due to the reasons rooted in the historical evolution of the community the region's economy has been transformed into the one over dependent upon remittances from outside working population. This has resulted in:

- Depleted farming hands.
- Resource mismanagement due to the lack of adequate manpower.
- Agricultural lands, specially far-flung ones, left barren for nobody's home to care for these.
- Depleting farm yields due to inadequate care and inputs.
- Depleting cattle count due to the lack of manpower to mend these.
- Depleting cattle productivity due to mismanagement induced by lack of manpower.
- Lack of innovation in farm sector for the left over population is not in a position to maintain what they have, so it does not find him/her psychologically prepared to take up more burden and to add to it the people do not have surplus to take risks.
- Increasing drudgery of the females who have to look after the various agricultural activities besides collecting water, fodder and fuelwood and also take care of the routine household chores.
- Lack of social cohesion due to fragmentation, out-migration and modernization.

- Breakdown of the traditional system of resource management due to neglect, improper documentation.
- Increasing malnutrition; specially amongst children and women.
- Rampant poverty.

Serious and concentrated efforts are needed to revert this trend that seems to have crept deep into the lives of the masses at grassroots level.

With the limitations of working hands and enough agricultural lands the region is well suited to horticultural operations and these have the potential of revitalizing the economy of the region besides improving the nutritional value of food available to the masses; particularly women and children.

Project Intervention Area

The Chamba-Mussoorie Horticultural Belt (*Falpatti*) that was conceived with a vision to boost horticultural production of the region wayback in 1965 has failed to yield desired results. Non-availability of quality planting material together with timely training and extension services are seen as the reasons for the failure of this scheme. The present project was conceived for revitalizing this belt and set forth a replicable model.

The various interventions of the proposed project revolved around Training-*cum*-Demonstration Farm at Kanatal that is being developed into a centre for the demonstration of new crop varieties and horticultural plants as also for undertaking training programmes. The list of the villages where interventions were undertaken during winter season is as follows: Under the Chamba-Mussoorie Falpatti where the winter fruit plantation is being done are San Gaon, Khuret, Thang Dhar, Raunsali Khal, Charipalda, Jaripani, Sour, Chopriyal Gaon, Nair, Kainthogi, Tikhon, etc. whereas on the Gangotri highway Harsil area is also being covered under which the villages, where the apple plantation is being done, are Mukhwa, Tharali, Sukhi Top, and Harsil. Under this plantation programme the Ghansali area which is located in the middle of the Chamba-Rudarprayag road is also being covered where the winter fruit plants, stone fruits and apple plants are planted in Duari villages which is 8 km away from the Ghansali road.

Under the Ranichauri area the villages are Sabli, Moun, Jagdhar, Tegana, Dharkot, Pali, and Gajana where stone and pome fruits are being planted in winter season.

Details of fruit plants distributed during winter season

(A)

Sl.No.	*Fruit Plants Name*	*Stone fruit*			
		Kanatal Area	*Ranichauri Area*	*Ghansali Area*	*Total Plants Distributed*
1.	Pear	3000	400	1000	4400
2.	Peach	700	295	500	1495
3.	Plum	100	–	100	200
4.	Apricot	1500	450	1000	2950
5.	Walnut grafted	–	133	–	133
6.	Walnut seedlings	2000	564	1500	4064
7.	Persimmon	–	–	200	200
8.	Kiwi	–	–	200	200
9.	Almond	1000	600	800	50
10.	Mix Peach and Apricot	–	50	–	50
	Total	**8300**	**2492**	**5300**	**16092**

(B)

Sl.No.	*Fruit Plants Name*	*Stone fruit–Apple*			
		Kanatal Area	*Ranichauri Area*	*Ghansali Area*	*Total Plants Distributed*
1.	Red Gold	2000	200	600	2800
2.	Tydomans Early	1000	200	200	1400
3.	Golden Delicious	600	–	–	600
4.	Starkrimson	1000	–	200	1200
5.	Red Chief	2000	100	400	2500
6.	Vance	4000	400	1000	5400
7.	Top Red	4000	200	500	4700
8.	Royal Delicious	8000	500	1500	10000
9.	Red Delicious	1100	–	–	1100
	Total	**23700**	**1600**	**4400**	**29700**
	Total Farmers	**319**	**85**	**112**	**516**

(C)

Sl.No.	Apple Distributed in Harsil Area	No. of Plant
1.	Red Gold	7500
2.	Tydomans Early	2000
3.	Golden Delicious	2000
4.	Starkrimson	3700
5.	Willium Spur	1800
6.	Manchuria	100
7.	Royal Delicious	28000
8.	Red Delicious	15000
Total		**60700**
Total Farmers		**602**

Total plants planted in winter season–1,06,492
Total Farmers –1118

1. Name of the Partner Farmer: Shri Kesav Dutt Bahuguna S/o Shri Ram Pd. Bahuguna
2. Address:
   - 2.1 Village: Ranichauri (Khov)
   - 2.2 Block: Chamba
   - 2.3 Contact number (if any) –
3. Socio-economic status of the farmer:
   - 3.1 Main profession: Government Service
   - 3.2 Familial landholding: 20 Nali
   - 3.3 Annual household income: 90,000
4. Details of the support extended to the farmer:
   - 4.1 Plants given to the farmer: No.
   - 4.2 Apple 24
   - 4.3 Peach 10
   - 4.4 Plum 10
   - 4.5 Apricot 10
   - 4.6. Walnut 10
   - 4.7. Cost of the plants in rupees 1152
   - 4.8 Horticultural training programmes attanded by the farmer: 01
5. I certify that the information furnished above is true to the best of my knowledge and I am extremely satisfied with the various interventions being carried out in the area by HIMCON with the support of Dewan Foundation, London. I am hopeful that these interventions lead to my prosperity and the overall properity of the region. I agree to pay the above mentioned cost of given plants at the time of first fruiting to HIMCON.

Signature of the beneficiary farmer

Kesav Dutt — Name of the beneficiary farmer

1.	Name of the Partner Farmer:		Shri Sanjay Bahuguna S/o Shri Chanderballav Bahuguna
2.		Address:	
3.	2.1.	Village:	Ranichauri
	2.2.	Block:	Chamba
	2.3.	Contact number (if any)	–
3.	Socio-economic status of the farmer:		
	3.1.	Main profession:	Agriculture
	3.2.	Familial landholding:	30 Nali
	3.3.	Annual household income:	50,000
4	Details of the support extended to the farmer:		
	4.1.	Plant given to the farmer:	No.
	4.2.	Apple	05
	4.3.	Plum	05
	4.4.	Almond	10
	4.5.	Apricot	10
	4.6.	Cost of the plants in rupees	540
	4.7.	Horticultural training programmes attended by the farmer:	01

5. I certify that the information furnished above is true to the best of my knowledge and I am extremely satisfied with the various interventions being carried out in the area by HIMCON with the support of Dewan Foundation, London. I am hopeful that these interventions lead to my prosperity and the overall prosperity of the region. I agree to pay the above-mentioned cost of given plants at the time of first fruiting to HIMCON.

Sd/-	Signature of the beneficiary farmer
Sanjay Bahuguna	Name of the beneficiary farmer

2 Dignity of Labour Camps

Agricultural Universities and Secondary Schools

Introduction and Objectives

A "Dignity of Labour Camp" was organized at the Mahatma Phule Krishi Vidyapeeth Central Campus, Rahuri, from 1 to 30 June, 1974 and continued from 8 July to the end of September 1974, with the object of :

(i) inculcating the dignity of labour in the younger generation as well as in older people;
(ii) developing and applying measures for land and water conservation and development, such as water harvesting, compost making by using farm wastes, and planting programmes for fuel, fodder and fruit trees, by using students power;
(iii) training of students in the above activities for large scale application; and
(iv) extending these methods for application in other educational institutions and in villages from where student participants come.

In the first fortnight of June, 555 students of whom 64 were college students, 341 agricultural school students and 150 non-resident students from neighbouring villages, participated. Two hundred and ninety-two students, of whom 52 were college students, 160 agricultural students and 80 non-resident students, took part in this camp in the second fortnight of June.

The camp was continued in July 1974 with the participation of 85 diploma holders of the Agriculture Schools. High School students of the IX standard from Ahmednagar city and Rahuri Taluka participated in this camp from 8 July for 2 to 4 days as a part of their social work week. A daily average of 60 students came from Ahmednagar city covering the entire strength of the IX standard from neighbouring high schools. They were given sprouted Makki, Cowpea and Gram at work sites at about 10 a.m. daily. They were also provided with free transport by the University.

From 8 July 1974, to the end of September of that year, 3,383 student days (2,680 boy days and 703 girl days) were utilized as part of the Dignity of Labour Work.

Achievements

The objectives essentially achieved were awareness of "Dignity of Labour" and education in the field of soil and water conservation, and land and water development, which was communicated to the participants in the field.

Still more specifically, the following have been the achievements:

(i) completion of a survey and designing of ten ponds;

(ii) a contour survey of part of the MPKV campus (2,000 acres);

(iii) construction of five ponds completed (ponds number 3 and 5 were partly done with the help of hired labour);

(iv) completion of twenty gully plugs:

(v) digging of compost pits and filling of 320 of these (6m×2m×0.80m), giving 19.8 m^3 of compost each, or a total of 3,450 m^3 of compost;

(vi) digging of pits, normally of a size of 60m×60cm×60cm in horticultrual areas (Dharmadi, hill 1 and 2), in the main orchard, and behind the Basic Science Building, also in marginal and sub-marginal lands of the Dairy Complex

1-2. School girls and boys collecting stones for terracing for soil conservation

3. Filling compost pits

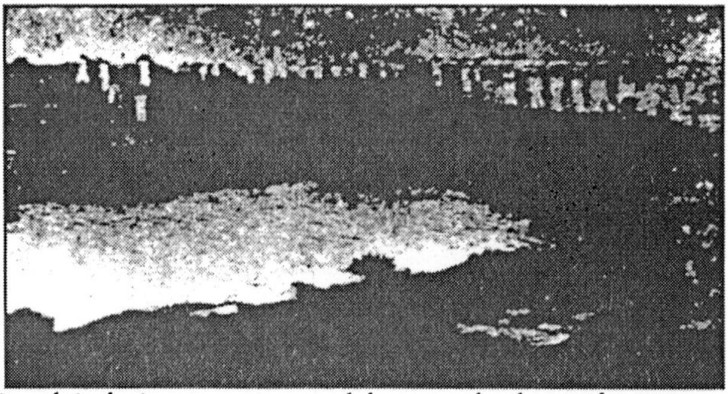

4. Pond is being constructed by youth through cooperative efforts

5. Yield programming for high per acre production of wheat on farmers' fields

6. Author (shirtless) working with school boys in preparation of compost pits

and other parts of the campus, especially the 'C' block; filling of these pits with soil and manure or compost; planting of total of about 50,000 trees (some with the help of hired labour) in pits; in some cases digging of water channels for irrigation of these trees;

(vii) picking of stones from an area of over 32 ha and putting these in trenches dug along contours (three ha of which have been planted with Marvel grass and 3.2 ha with safflower);

(viii) uprooting of Parthenium weed from an area of 600 ha (partly done by staff and students);

(ix) planting of 6 ha of the cleared land with jowar, and giving about two life saving irrigations from ponds (Plate-4) seventy quintals of jowar were harvested from the above area, which are to be distributed to farmers as "seed";

(x) planting of about one ha of agave bulbs which were making good growth and were expected to serve as a nursery for the coming season;

(xi) utilization of more than 300 cactus bushes for planting Neem seed through the "Bamboo" spear device; by this means the young seedlings were protected from stray cattle and they were provided with moisture by the cactus bushes.

Expenses

The total amount spent in June, July, August and September, 1974, for these items was Rs. 94,584.30

		Rs.
(i)	Remuneration to students	47,308.75
(ii)	Meal charges for students	33,237.50
(iii)	Fuel and oil for buses, jeeps and cars	10,038.05
(iv)	Repair to vehicles	4,000.00
	Total :	94,584.30

Achievements of Agriculture College, Dhulia

At our instance, a "Dignity of Labour Camp" was also organized by Dr. B.D. Khot, Associate Dean, Agriculture College, Dhulia, from 27 October to 1 November, 1974, in which 400 students from different High Schools participated. The work in which students were trained were compost making, calf and poultry house management, harvesting of sunflower, raising of nursery plants and seedlings, cleaning and laying out of play-grounds, etc.

Camps at Agricultural Schools of MPKV

All the nine agricultural schools attached to MPKV have in their curriculum "work experience" as an item of work. All farm operations are done by the students. Actually a stipend of Rs. 40.00 p.m. is paid to each student as maintenance allowance. However, some special activities such as compost pit preparation were undertaken by most of the schools and a standard rate of payment was made to the students for each compost pit prepared. The camp thus became an "earn while you learn" exercise.

Camp at Sholapur

Looking at the progress at the central campus of Rahuri and also at Dhulia, Dr. N.D. Patil, Chief Scientist, All India Coordinated Research Project for Dryland Agriculture, organized a similar camp at Sholapur for three days.

Large-Scale Involvement of Secondary School Students

Proposals were under way for large scale involvement of students for a period of 1-2 months per year in works of a community nature such as :

(i) Digging of pits and planting of trees;

(ii) Maintenance of planted trees;
(iii) Compost pit digging and filling;
(iv) Water harvesting and pond construction;
(v) Other tasks as and when required by the neighbouring villages.

It is proposed that each high school and college adopt a village. The Centre for Rural Studies, associated with Ahmednagar College, is doing a remarkable job of such adoption and helping to revitalize the rural scene.

The task of the University is to train selected teachers of schools, so that they may serve as stimulants for their students.

The experience gained in the summer and monsoon of 1974 showed that there was a great variability in the performance of each batch of students depending considerably on the ability, devotion and efficiency of the teachers accompanying the batch.

Collaboration between MPKV and Maharashtra State Education Department was stimulated and regular liaison was established between the two organizations. I am sure there is great scope for such collaboration between similar organizations in the field of rural and even urban development in India, utilizing the energy of young students for national development.

3 Soil and Land Resources Evaluation

Introduction

The present status of soil and land resources study in Maharashtra is as follows:

(i) The State Soil Survey Department under the Department of Agriculture is charged with the task of surveying soils in various parts of Maharashtra. Each district has a soil survey party consisting of a Soil Survey Assistant plus Field Assistants. The soil survey is done on the basis of the standard *Soil Survey Manual* developed for India by the All India Soil and Land Use Survey Organization. The village maps on a 8^2 to a mile or 1: 7,960 scale are used as base maps. The soil series are mapped and symbols are put in each mapping unit essentially consisting of soil texture – depth, slope, erosion and a land capability class.

(ii) The Regional Soil Correlator located at Nagpur and under the administration of the All India Soil and Land Use Survey Organization (now being developed as the All India Bureau of Soil Survey) takes care of soil correlation work plus special soil surveys of stations, universities, specific projects, etc. The same procedure is adopted as above. The aerial photos, although taken are not readily available for reasons for security.

In spite of some familiarization and consultations with the chiefs of both the above groups, it was felt that with the short time available, I should spend it on soil and land evaluation and interpretation for Agricultural Development Programmes. In this context, I took three sample areas:

(i) The Central Campus of about 8,000 acres (3,200 ha) at Rahuri for which I prepared a detailed plan for land, water and crop development (on a 1:8,000 scale, approximately) (See Chapter on Master Plan)
(ii) Rahuri taluka - total : 1,01,685 ha
(iii) The Ahmednagar district - total : 17,02,039 ha

Where some information on soil series was available and additional information was obtained, land evaluation maps and data were prepared. The maps and data for Ahmednagar district and Rahuri were also prepared.

Land Resources Evaluation

The land resources evaluation of a region is a comprehensive view of the inventory of its land resources, eventually supplemented by a multipurpose evaluation of its potentialities. The Land Resources Inventory has two major aspects:

(i) A Land Resources Map, having a legend with land units which are defined and described on the map; and
(ii) An evaluation of the potentialities of each of the land units.

A land unit is defined as a given combination of characteristics of topography, climate, soils, geology, vegetation, hydrology and land use. The mapping of these land units provides one with a land resources inventory.

The first stage is on the basis of the physiographic land type into which a region or a district is sub-divided. A land type is defined on the basis of physiography and the nature, thickness and continuity of the soil cover, using the natural relationship between the land forms and the soil cover. The advantage of making these groupings is that they have also significant land uses. For example, in Ahmednagar district, the following physiographic land types have been defined:

(i) Flat higher plateau (Mesa)
(ii) Hill slopes
(iii) Upper Piedmont
   (a) Piedmont back slopes
   (b) Piedmont mid-slope or undulating plateau
(iv) Lower Piedmont – Piedmont toe slopes
(v) Flood plain
   (a) Inner terraces
   (b) Outer terraces
   (c) Levees
(vi) Complex land type
(vii) Miscellaneous rock outcrops
   (a) Complex land type
   (b) Miscellaneous rock outcrops.

Preparation of Land Resources Inventory

This is done by compiling in stages all the available material on the area or region including reports and maps, topography, geography, geology, land classification, water resources, land use etc. The useful material is then sorted out. To fill in any knowledge gaps, some fieldwork is done. Subsequently the field map is compiled, in this case on a 1:125,000 scale, followed by interpretation, evaluation and presentation of land potentiality. In all the above, Mr. Lande, Assistant Soil Survey Officer from the Soil Correlation Centre, Nagpur, assisted me through the kind courtesy of Dr. R.S. Murthy, Chief Soil Survey Officer, New Delhi,* and Mr. Bhattacharya, Soii Correlator, Nagpur.

Land Resources Potentiality

Having given the land resource unit or type, it is necessary to give the potentiality for a specific land unit.

The general definitions of land potentiality apply for each of the major types of land use, namely :

- irrigated annual/perennial crops
- dry farming /rainfed farming
- range
- forest

Land potentiality classes are defined in two different maps:

(i) without initial land improvement and/or land conservation works;
(ii) after initial land improvement and/or land conservation works.

In the first case, the land class gives an assessment of the *present* potential, that is the potentiality of the land at the level of the management considered and with the present limitation of the land. In the second case, the land potentiality class indicates the potential *after* the initial input which is used for removing some of the limitations and at the level of the management considered.

Eight classes of land potentialities are defined, just as in the Land Capability classification, from class I, high potential, to class

* Presently Director, National Bureau of Soil Survey and Land Use Planning, Nagpur.

VIII, very low potential. Inputs A to D are defined as A – low to D – very high. Future potential can be given after the improvement has been made.

Geomorphology and Soils In Ahmednagar District (M.S.)

Sl. No.	*Geomorphic Units*	*Soil Series*	
1	2	3	4
1.	Flat higher plateau (Mesa)	Sirasgaon	Srg
		Wadgaon	Wdg
2.	Hill slopes	Sibneri	Sib
		Sindudi	Snd
	Piedmont Plateau		
3.	Upper Piedmont		
	(i) Piedmont black slope	Khanpur	Knp
	(ii) Piedmont mid-slope or undulating plateau	Telegaon	Tlg
		Pargaon	Prg
		Torkewadi	Tkd
4.	Lower Piedmont or Piedmont toe slop	Sawargaon	Srg
		Nimone	Nim
		Annapur	Anp
	Flood Plain		
5.	Inner terraces	Umraj	Umj
		Eklahare	Ekr
6.	Outer terraces	Otur	Otr
7.	Levees	Dholwad	Dld

Remote Sensing Technology for Land Resources Evaluation of Maharashtra State

Introduction

1. The Government of India has recently set up a National Remote Sensing Agency (NRSA) and has embarked on initial projects at the national and state levels, involving rapid computer analysis for the evaluation of natural resources.
2. Modern remote sensing capability has made it possible to rapidly and inexpensively assess the soil, climatic, water

and overall natural resources of large areas such as a state, or the country as a whole, in order to derive computer based optimization models and feasibility studies for their overall development, and subsequently guide developmental projects on a sound physical and techno-economic base.

3. Projects of this nature have already been initiated for a first reconnaissance level assessment for natural resources of various states and the centre, through the National Remote Sensing Agency set up under the Department of Science and Technology. The Government of Maharashtra is also considering the utilization of this new technology for undertaking studies for the following objectives to be achieved.

Objectives

General Long-term Objective

Physical maps are the base for all concrete planning. The following maps would be provided along with detailed tables for the State of Maharashtra:

(a) Climatic, physiographic and geographic maps, at the reconnaissance level;
(b) Tectonics, geomorphic, geology and potential mineral maps, including prospective oil sites;
(c) Soil maps giving 4 to 6 main groups and sub-groups;
(d) Land use capability maps for the entire State, with 6-8 land use capability classes;
(e) Current land use, including forests, range lands, scrub land, waste land, and cropped land (8-12 classes);
(f) Hydrology and water resources; and
(g) Urban land use map for the Bombay Metropolitan region.

The long-term objectives of obtaining such evaluation would include the next step of drawing up feasibility projects for development in the areas of soil resources, forest resources, agricultural development programmes, water resources development, particularly for the entire Western Ghats zone of Maharashtra, development of fertilization and manuring of soils, overall rural development particularly of the drought prone areas, and the development of the rural and urban complex to serve the problems of the State as a whole. Subsequently, projects in each area of socio-economic activity would be derived on the basis of optimization models with the objectives of the greatest benefits in terms of economic returns on investment, optimal employment, redistribution of wealth, and any other desired goals of the Government of Maharashtra in the overall framework of the development activities of the country as a whole. Furthermore, management models for optimal resources utilization and conservation would be derived for the Departments of the Government concerned, to be guided by in the overall process of planned economic and social development of the State as a whole.

Specific Objectives

The specific short-term objectives of the Remote Sensing Project for natural resources assessment would be as follows:

(a) to acquire maximum information on land use, forest, range land, agricultural crop resources by seasons, surface and ground water resources and soils of Maharashtra covering an approximate area of 3,00,000 km^2;

(b) to prepare land use capability maps of the State of Maharashtra;

(c) to render current land use maps for the *kharif* and the *rabi* season for the State;

(d) to prepare the hydrogeology and water resources maps by catchments and sub-catchments for the entire State;

(e) to demarcate problem areas with regard to the intensity and extent of droughts and with respect to severe erosion, salinity and alkalinity, etc.;

(f) to prepare the surface drainage pattern of each water-shed in the State of Maharashtra;

(g) to prepare tectonics and geomorphology maps with prospective groundwater and mineralization sites for the State including an initial physiographic survey of the continental shelf of Maharashtra; and

(h) to prepare an urban land use survey for the Bombay Metropilitan area:

Maps for the above categories would be rendered at the following scales:

1 : 1,000,000	(orthophotomosaics)
1 : 250,000	(orthophotomosaics)
1 : 50,000	(graphic plotter)
1 : 25,000	(L-Map Print-out)

Detailed Output Results to be Obtained

1. To cover the entire 300,000 km^2 (approx.) with black and white and colour mosaics (semi-controlled) of Maharashtra State, referred to above. Scale: 1:1,000,000 (corrected for earth's rotation, pitch, roll, etc. with UTM grid).
2. To cover 100,000 km^2 area of the drought prone areas of Maharashtra for existing land use and capabilities and water resources (output in EBR or LBR in colour for three most relevant themes; corrected maps on scale of 1:250,000-1,330,000).
3. Over 100,000 km^2 (equivalent to approximately 10 ERTS frames) of selected target areas for geology, hydrogeology groundwater, and minerals prospective sites.
4. Output in (a) 1: 50,000 scale corrected, grey-scale plotter print-out for 100,000 km^2 of the defined target areas;

(b) 1: 24,000 scale on L-map print-outs uncorrected for any three chosen themes from above.

5. To obtain the detailed tectonics, geomorphic and geological interpretation of the target areas defined above through enhancement technique interactive computer analysis in the following semi-detailed scale print-outs:

   (i) graphic plotter grey-scale map on 1:63,360 or 1: 50,000 for any one land use, or other theme with 6 to 10 classes (corrected) or 60,000 km^2 of target areas;

   (ii) L-map print-outs in 1: 24,000 scale with multi-theme results in 6-10 thematic classes (uncorrected) of 100,000 km^2 of chosen target areas.

6. Tables of categorization of areas in each class for each time of the above would also be given.

A soils map of Maharashtra State exists on a 1:1,000,000 scale. It appears that this has been prepared from the existing maps and so no detailed checking in the field was done. A proposal for the use of remote sensing technology for soil and land use survey of Maharashtra has been made. This actually forms a part of the Maharashtra Institute of Natural Resources for Agriculture (MINRA). It is hoped that the following maps will be prepared through MINRA for Maharashtra in the next few years on suitable scales (ultimately 1:50,000 scale) : soil map, land resources potentiality map, land capability map, irrigability map, and present land use map.

4 Organics–Rural and Urban Composting

Introduction

In the present system of intensive farming the application of fertilizers plays a key role, not only in maintaining but also in increasing soil fertility, and soil productivity with its impact on the yield maximization. But the inorganic chemical fertilizers have become prohibitively costly and occasionally are also scarce. In addition to this, the hunger of tropical soils for organic matter is very intense. With this view, the use of organic manures and the recycling of organic wastes is becoming increasingly urgent and important. It is estimated that less than ten per cent of the total potential of compostable waste material is being exploited in Maharashtra and in India, while the bulk of it is being wasted.

A qualitative and quantitative potential has been estimated in a paper on "Rural and Urban Composting", presented to a joint seminar of the Fertilizer Association of India and FAO, where I was invited as a guest speaker to present a paper. This paper is reproduced in Appendix I.

For harnessing the natural resources of plants and animals, it is necessary to begin a campaign of "Use of organics in agriculture". Although composting has been well known in India for centuries, it is necessary to replace the old methodology by the new technology so far available. This would result in the partial replacement of the chemical fertilizers. As such a compost promotion programme is becoming urgent, the following action is suggested:

1. Oraganized training of the farmers and extension worker explaining the principles and available technology of composting.
2. Compost making on a large scale:
   (a) inducing the big as well as the small farmers in this vital activity;
   (b) inducing the Gram Panchayats, Municipalities, etc. to take sustained action; and
   (c) erection and operation of high speed large or small mechanical/semi-mechanical compost plants;
3. Establishment and running of *gobar* gas (biogas) plants on a mass scale.

Organized Training of the Farmers and Extension Workers, Explaining the Principles and Available Technology of Composting

In order to emphasize the importance of organics in building up the soil fertility and productivity, organized training programmes are necessary through which the farmers, extension workers and other agencies involved in agriculture, can be fully exposed to scientific methods of composting. A compost training class was organized at the Central Campus, Mahatma Phule Krishi Vidyapeeth, Rahuri, in which 30 extension workers participated. Prof. G.K. Zende, Dr. Daftardar, Dr. N.K. Savant and the writer explained the recent techniques of composting farm waste.

Large Scale Composting

During the conducting of the 'Dignity of Labour Camps' referred to earlier, the vital importance of organic manures in crop production was explained and demonstrated to the school and college students by Mr. J.H. Dongale, and the work of filling the compost pits was also undertaken. Three hundred and sixty-one pits (20×6×3 ft) were filled at the various sites of the Central

Campus, MPKV, Rahuri, including the dairy complex. The main raw materials used were wheat-straw and other waste materials such as groundnut husks, etc. plus cowdung of which slurry was made. Reinforcement was done by urea and superphosphate.

Mechanical Composting

The modern mechanical high-rate composting has resulted in the reduction of space and time for composting and as such, mechanical composting has become a proposition worth considering from the economical and commercial point of view. The compost prepared by the mechanical composting method contains more plant nutrients than that produced by the conventional method, requires less time (5-10 days), less space and also provide a means of treatment of sewage, sullage, night soil, ash and earth. There are, for example, about 160 DANO Composting Plants in various parts of the world which prepare the fresh DANO compost of better quality within three to four days.

So, the possibility of installing such high speed DANO Compost Plants (or other similar ones) at big farms such as the cooperative sugarcane farms as well as municipal areas, needs to be explored and encouraged. Semi-mechanical compost plants for small community solid waste disposal need to be studied and promoted.

A country like Japan, producing millions of tonnes of urea, has promoted in a big way the use of organics for rice and other crops. Although organics suffered a set-back due to the easy and cheap availability of chemical fertilizers in the sixties, since 1973, when the world prices of crude oil were raised, their value has been recognized.

A mechanical compost plant has been developed at the National Environment Engineering Research Institute, Nagpur, and this needs to be followed up immediately.

The DANO system of composting, as has been proved, has a future.

Gobar Gas Plant (Biogas)

Due to the acute shortage of fuel and fertilizers and the low purchasing power of the bulk of the rural community, the installation of cowdung gas plants, generally *gobar* gas plants, is becoming more and more important.

In this plant, cattle and other dung is allowed to ferment for a few days under anaerobic conditions, whereby a mixture of gases, consisting mainly of methane, hydrogen and carbon dioxide is produced. Methane and hydrogen can be used as a fuel and the residual matter, which is rich in plant nutrients, can be used as manure.

Considering its importance and multiphase utility, it was decided to install 100 *gobar* gas plants around and about the Central Campus during 1975. Prof. M.V. Chavan, Extension Planning, was to extend the *gobar* gas plant ideas to the villages, while Dr. S.D. Sawant, Agricultural Engineer, would furnish the technical aspect from the engineering point of view. The Khadi and Village Industries Board of the Nagar district, as also the State and Central Banks, were ready to give the financial aid to such family-size plants.

The installation, and also the functioning, of two *gobar* gas plants, one at the Dairy Complex, Central Campus, Rahuri, and another at Sade village, had already started.

Some details of the biogas plants and their operation are given in the references.

Four-Pronged Attack on Reinforcing the Organics Programme in Maharashtra

In addition to the above, suggestions are as follows:

(a) Increased use of water hyacinth in composting. Actual water hyacinth is an important and rich source of plant food and should be fully utilized (percentage of N is about 0.4, of P_2O_5 about 0.5, and of K_2O 0.6*).

* Variable.

(b) Glyricidia planting on marginal and sub-marginal lands should be promoted as it is being done throughout the MPKV establishment. These plants should be planted in millions on the sub-marginal lands and compost centres be developed in each village or at the most for a cluster of villages.
(c) Sesbania planting could be used for fodder and also for composting.
(d) Fish meal composting could be taken up in coastal areas, as these resources are tremendous. Konkan Krishi Vidyapeeth should play an important role in this programme.

5 Utilizing Energy of Youth for National Development

Introduction

While great emphasis is being given to soil, water and other natural resources development, *India has to pay more attention to her greatest asset,* viz. *human energy.* The period of maximum energy and vigour is between the age of 15 to 25. It has been estimated that over *120 million youths* of both sexes fall into this age category. The utilization of this *vast resource of energy for natural reform development is at a low level of efficiency and must be improved.*

Great emphasis was given to contact with youth groups and participation in their activities, leading to an understanding of their thinking to see what action programmes can be initiated leading finally to specific agricultural and national development.

Exploratory or Pilot Programmes

The following specific steps were taken in this direction:

(i) Initiated Dignity of Labour camps and greater contact with the youth – thus, worked with and came in almost *daily contact with over 2,000 youths at the Central Campus, Rahuri;*

(ii) Addressed 1,000 National Cadet Corps (NCC) in Poona and spent a day with them in their camp. Topic of address *"Youth Corps in Iran with some suggestions for India";*

(iii) Discussed with 30 NCC officers from different parts of the State, their views on utilization of energy of youth for national development;

(iv) Addressed 1,000 NCC cadets at Ahmednagar and discussed with 15 officers the *problems and potential of the NCC;*

(v) Familiarised myself with the working of the National Service Scheme through the training progammes being conducted at the Centre for Rural Development, Ahmednagar College, Ahmednagar by Prof. Hulbe, Director of the Centre, and his group, including the village adoption programme;

(vi) Helped in the *State Level Youth Festival Organized by the Maharashtra Youth Centre* (MYC) with 500 participants from all parts of the State including students and non-student youth, boys and girls, rural and urban youth. It was a three-day camp wherein I participated specially in organizing:

   (a) *A Dignity of Labour Work,*
   (b) *An Exhibition,*
   (c) *Seminar on "Role of Youth in National Development".*

Topics discussed in the seminar were:

Rural Development

1. Agriculture, Forestry, Fisheries, Dairy Development,
2. Rutal unemployment and underemployment,
3. Multiple purpose rural development:

   (a) *health and sanitation,*
   (b) *agro-industries,*
   (c) *education,*
   (d) *family planning.*

Urban Development

1. Urban unemployment and underemployment,

2. Development of urban industries and their impact on the rural area.
3. Use of urban youth for rural development.
4. Family planning.

Socio-economic and political structures in India and their impressions on the minds of youths.
The exhibition organized at MYC was in two parts:

(1) Natural Resources of Maharashtra.
(2) Dignity of Labour Camps.

The MPKV/AHMEDNAGAR Youth Centre part of the exhibition was the prize winner.

(vii) Preparation of paper on "Youth Corps – a Blue Print for India" prepared together with my son, Vikram Dewan, of the International Development Centre of Cornell University, Ithaca, New York, USA, together with an open letter addressed to all interested in utilizing the energy of youth for National Development;

(viii) A project prepared by my son Vikram and other colleagues in the University on "Students Training and Orientation for Leadership in Accelerated Agricultural Development" – STOLLAD (Appendix III);

(ix) *A Pilot project for "Agriculture Oriented Work Experience"* was prepared for the secondary schools together with the Director of Education, M.S., Dr. Chitra Naik. This is for a one-week work experience programme for all the students of Grades VIII and IX, normally between the ages of 14-16 years.

Objectives

1. The objectives of such a project (which is hoped to be applicable to a majority of secondary school students of

these 2 classes and more in Maharashtra State – a total of about 2 million) are:

(a) *To associate young students in agricultural programmes by seeing and participating in these programmes so as to make them agriculture-oriented.*
(b) To involve all the students specially with a rural background in the basic operations of agriculture including new varieties of crops, fertilizers, manures, irrigation, crop protection, harvesting and post-harvest technology.
(c) To attempt basic improvement in agriculture specially for marginal and sub-marginal lands, specially in the water harvesting and tree planting programmes.
(d) To help in environment improvement and *preparation of organics and compost and waste material recycling.*

2. In the first stage, all the schools within 40 km distance, of the following will be involved:

(i) Central Campus, Rahuri;
(ii) Agriculture College, Kolhapur;
(iii) Agriculture College, Dhulia; and
(iv) Agriculture College, Poona.

3. It is expected that about 5 × 10,000 students will participate at each campus. *The Education Department will arrange for the transport of the students. The university or college will arrange for feeding the students and their lodging.* An area would be allocated for crops and vegetable production at each campus and the workforce would essentially be provided by the students. This programme could also be extended to *important Research Stations such as the Sholapur Dryland Farming Station.*
4. This is a joint project between the State Education

Department and the Agriculture University for providing this "*Agriculture Oriented" experience* to young High School students in large numbers through which the new agriculture technology will be spread. It is hoped that all the Agricultural Universities will follow such a joint programme. An emphasis will be given to training of teachers of Secondary Schools – mainly teachers in charge of "*Work Experience*" classes. Utilization of "*Drop-out students**" should form a large project for rural and agricultural development.

* "Drop-outs" are 110 million or 11 crores for the whole of India and about 10 million for Maharashtra State.

6 Farm and Social Forestry

Introduction

In many parts of Maharashtra, especially in the semi-arid tracts there is little and often no useful vegetation. The estate of the Mahatma Phule Krishi Vidyapeeth including the seven adopted villages, viz. Rahuri Kh., Rahuri Bk., Digras, Warwandi, Pimpri Avghad, Sade and Khadamba Kh., have considerable areas of land, with little or hardly any natural vegetation and the land is often badly eroded. So there is great need for farm forestry, which is also called social forestry, to fulfil the following objectives.

Objectives

1. To afforest and plant trees on the marginal and sub-marginal hills soil and slopes for the conservation of soil and moisture.
2. To beautify the university campus by planting ornamental and flowering trees.
3. To develop the economic status of the area by planting tree species having fodder, fuel and other economic value.

With the above objectives in view the following necessary tasks have been/are being started:

1. Raising of farm wood lot.
2. Afforestation of village waste or idle land.
3. Wind breaks which are very much needed in areas like Rahuri.

4. Shelter belts.
5. Afforestation of canal banks and irrigation channels.
6. Tree planting on roadsides.
7. Afforestation of receding water areas for major, medium and minor reservoirs.
8. Forestry for reclamation—waterlogged non-irrigable areas, hill tops etc.
9. Urban tree planting.
10. Recreational forestry.

Farm and Social Forestry Nursery

Sl.No.	*Scientific Name*	*Local Name*
1.	*Azadicchta indica*	Neem
2.	*Dalbergia sissoo*	Sizoo
3.	*Acacia arabica*	Babul
4.	*Albizzia lebbek*	Siris
5.	*Tectona grandis*	Sagwan
6.	*Agave sisaliana*	Sisal
7.	*Acacia catechu*	Khair
8.	*Pongamia Pinnata*	Karanj
9.	*Hardwidai binata*	Anjan
10.	*Santalum Album*	Chandan
11.	*Cassia siamea*	Kashid
12.	*Tamarindus indico*	Chinch
13.	*Casurina equisetifolia*	Suru
14.	*Dendracalamaus strictus*	Bamboo
15.	*Eucalyptus globulus*	Eucalyptus
16.	*Punica granatum*	Pomegranate
17.	*Psidium guajava*	Guava
18.	*Citrus aurantifolia*	Kagazi lime
19.	*Annona squamosa*	Annona
20.	*Grewia asiatica*	Phalsa
21.	*Zizyphus mauritiana*	Ber
22.	*Ficus carica*	Fig
23.	*Mangifera indica*	Mango
24.	*Carica papaya*	Papaya
25.	*Musa paradisiacal*	Banana

Nursery

With the consultation of Mr. C.P. Bhimaya and Mr. V.D. Mahendale, retired Forest Officers, on the above objectives and tasks, the nursery works have been started. In this connection 5 ha area of the horticultural nursery in the 'C' block of the farm has been allocated. Preparatory tillage has been completed and also the sowing of the following tree species is either completed, or is in progress or is planted.

The nursery is organized as a series of raised beds about 1metre wide, 2.5 metre long and about 10 cm high. The water channels are provided so that the beds can always have ample water but are never waterlogged. Weeding must be done regularly. The plants can be put in plastic bags with soil and transplanted as required after removing the plastic bag. It is cheaper to transplant them after taking them directly from the beds.

Tree Plantation

To encourage the tree plantation task, during the Dignity of Labour Camps held at Rahuri, during June-September 1974, several pits and trenches were dug and about 50,000 trees planted on Dharmadi hills, at the dairy site and other parts of the campus. These trees are being maintained, partly by gravity irrigation, partly by ooze irrigation (see Appendix V) and are growing with good vigour with a survival rate of over 95 per cent. The recommendations of the retired Forest Officers suggest the prominent species of trees for various purposes, such as planting on canal banks, on eroded hill sides, on road sides, etc.

Regarding the seven species of hardy fruit trees (excepting Papaya and Banana) they can be successfully planted on hill tops and hill slopes with poor soils. The pits of 60×60×60 cm or more can be dug, filled with good silt and clay and also a basketful of FYM or compost and a half kilogram of single super-phosphate. Mango can be planted at the base of the hill where the soil is much better than the hill slopes and hill tops.

In between, mango trees, *phalsa* can be planted so that the land between the mango plantations can be gainfully utilized.

Papaya and banana fruit trees can be planted on the medium

type of soils. Papaya can be planted along the water-channels all over the four boundaries of the field as also along the farm road.

The preparation of area for tree planting has to be done with care. Some details of trees and plants are given in Appendix IV.

Plantation of Trees at an Advanced Age

As I was familiar with the technology of the transplantation of mature trees in Europe, an effort was made to transplant over 10 years old Neem tree from the Dharmadi to the campus area near the quarters which looked so dreary and dry. Four Neem trees were transplanted in holes about 2 metres long, 2 metres wide and about 2 metres in depth. The tree, with the soil and roots, as far as possible were brought to the site to be transplanted. Transplantation was done before the monsoon. The leaves were removed from the tree, so as to reduce evaporation. Fortunately an early shower (27 May, 1974) came at the time of transplantation.

One of the four trees have survived. During the coming monsoon, more mature trees will be transplanted and the survival rate may be increased by planting in the middle of the monsoon.

7 Additional Resource Promotional Activities

Horticulture, Solar Energy and Waste Recycling

The cuttings of two horticultural crops, i.e. fig and pomegranate of ten varieties were received through the courtesy of the Government of Iran, Ministry of Agriculture and the Indian Embassy, Tehran in March 1975. Also the vegetable seeds of twelve varieties (each 30 gms) from M/s. Ferry Morse Seed Company (Inc.) P.O. Box No. 100, Mountain View, California-94042, USA were received through the FAO, Rome at my request. Earlier twelve vegetable varieties from Italy were received in January 1975. In August 1974, walnuts, pistachios and almonds from Iran were also planted both in Rahuri and in Mahabaleshwar. Transplanting in the field was done in April 1975. Pyrethrum was also planted in Mahabaleshwar on 1st April, 1975. All the cuttings and the seeds were handed over to Dr. A.V. Patil, Head, Department of Horticulture, MPKV, Rahuri for their planting and subsequent testing under our conditions. The details regarding the varieties are given below:

Fruit Crops

Figs	*Variety*	*No. of Cuttings*
1.	Onion Colour	20
2.	Yellow Red	20
3.	Yellow	20
4.	Black	13
5.	Unidentified	20
		93

Pomegranate	*Variety*	*No. of Cuttings*
1.	Agha	10
2.	Tabest	10
3.	Almo	10
4.	Siah-Shirin	10
5.	Dorshi-Macs	10
		50

Vegetable crops

FAO seed accession number	*Species*	*Cultivators/Varieties or common/vernacular name*	*Weight of sample*
35.915	*Lycopersicum esculentum*	Mech. Harv. VF 145-F5	30 gms
35.916	*Lycopersicum esculentum*	Red Cherry Laggs	30 gms
35.917	Lycopersicum esculentum	Tropi-Gro	30 gms
35.918	*Asparagus Officinalis*	Mary Washington 500W	30 gms
35.919	*Spinacea oleracea*	Long Standing Bloomsdale	30 gms
35.920	*Cucurbita pepo*	Jack O'Lantern	30 gms
35.921	*Cucurbita pepo*	Large Yellow or Connecticut Field	30 gms
35.922	*Cucurbita pepo*	Var. Melopepo Cozini	30 gms
35.923	*Hibiscus esculentus*	Emerald	30 gms
35.924	*Brassica fimbriata*	Dwarf blue curled	30 gms
35.925	*Brassica pekinensis*	Michihli	30 gms
35.926	*Brassica Oleracea Acephala*	Georgia Southern	30 gms

Promotion of Greater Use of Solar Energy and Wind Energy in Agriculture

There is plenty of sunshine in India and therefore, abundant solar energy is available. The energy on an average day is equivalent to 6-8 kilowatt hours per square metre. For normal plant photosynthesis, we are making use of 1/1,000th of the energy that is available to us. We have also a lot of wind energy available specially in some months of the year. We must develop and also grasp the available technology for utilizing the solar and wind energy.

Prof. Sanyal of the Agricultural Engineering Faculty of MPKV, is working on wind mill as well as additional technical information to utilise wind energy. Dr. S.D. Sawant and Prof. M.C. Varshney of the same Faculty are collecting information on solar energy related to solar water heaters, a solar dryer for small farmers and also a solar energy collector etc. and are working on some research and development plant.

For the above aspects of energy utilization considerable references, discussions and guidance was given to the staff concerned.

Parthenium from the fields and utilization for Card-Boards

Parthenium hysterophorus linn is an introduced obnoxious weed multiplying very rapidly in the State. It is attacking the fields, pastures, rivers and canal banks, and other areas of Maharashtra thereby reducing the productivity of the land. A campaign to check its expansion and to eventually eradicate needs to be initiated with great urgency. In this connection, the following proposals were made :

1. Experiments performed by Shri K.J. Soman, Director Handmade Paper Institute, Khadi & Village Industries, showed that card-board of good quality can be made from Parthenium digestion.
2. It is, therefore, possible to organize a campaign of uprooting Parthenium plants and using them in the manufacture of straw-board.
3. Workers, students and non-students engaged in the collection of parthenium may be paid a sum for every kg of uprooted plants.
4. The heads of the various educational institutes have to be approached to agree to mobilise a student force to work on the project. The MPKV is keen on participating in the project. The non-student population may also respond to the campaign.
5. We have requested the President of the Zilla Parishad and

Collector of Ahmednagar for full participation in this programme.

6. The best time for uprooting parthenium is before its flowering, but any time is alright to begin such a campaign. To boost this activity at the Central Campus farm and adopted villages, the students and staff helped to eradicate the parthenium in B-2 block on 15 August, 1974. The farm labour, technical and clerical staff also tried to eradicate the parthenium from part of the Warwandi village. The high school boys eradicated it from the pond site area of the Central Campus, MPKV, Rahuri. Some efforts were also made at the constituent colleges and Research Centres.

   The suitability of parthenium weed for the production of handmade paper and card-boards was also studied and the results are essentially positive as reported by Mr. K.J. Soman, Director of Handmade Paper Institute, Pune, who is deeply interested in this project.

Promotion of Use of Waste Material for Paper Production

Considerable promotion work was done for the use of bagasse; the waste material from the sugarcane industry, for paper. Presently, all or most of it is being used as fuel in the industry. However, if the furnace is efficiently designed, the quantity of bagasse in a factory is more than its fuel requirements. Additional bagasse can then be directed to the paper industry. One such plant has been started at Pravaranagar in Ahmednagar district.

To this raw material of bagasse, other materials such as Joncus, wheat straw, cotton stalks, even parthenium weed can be added for paper production.

It is recommended that the management of Sugar Factories should get together to discuss as to how best they can use their waste material and recycle it for the paper industry.

Planting of Trees in Areas on the sides of Railway Lines

Introduction

India has a large network of railways. The total represents over 60,000 km as follows throughout the country.

Name of the Railway	*Total Route km*
Central	6,012
Eastern	4,229
Northern	10,686
North-Eastern	4,977
North East Frontier	3,628
South Central	6,175
Southern	7,452
South-Eastern	6,842
Western	10,147

Area-wise, it occupies including 30 metres on both sides of the tracks, possibly 2,00,000 hectares or more. Together with the area of the yards, workshops and railway colonies, the total area with the railways would be more than 3,00,900 ha. Part of it, specially in the colonies, is planted with trees, bushes, creepers etc. Although, the whole area is not available for planting but there are large areas where suitable species of trees can be planted.

It is proposed that a project be worked out whereby all such areas be planted in a period of 10 years. My proposal for the first three years is as follows:

1. To survey the whole area, even with the use of satellite imagery on the lands which are covered by railway lines and in such cases establish how much area is with the railway on each side of the line.
2. To prepare a project plan for planting in selected areas on one or both sides of the railway lines.
3. To develop adequate number of nurseries for such a planting programme.
4. To get some railway staff trained in this important work.

Species of the Trees to be Planted

These will vary according to the region of the country. Rainfall is the important factor in every case but it is also observed that on either side of the railway line there are certain depressional or low lying areas which accumulate runoff water during the rainy season and retain water for a much longer time after the rains. These areas have greater soil moisture possibilities. However, the problems of production of these trees are large. No single solution can be given for such production but I feel that the forestry department can be of help especially as regards live fencing and ditch fencing. Certain species which are quick growing and serve as both fuel and fodder, as well as economic species of trees such as bamboo for paper production plus some fruit trees such as Papaya, Lemon and Guava, could be considered.

Organization

It is proposed that the Railway Board should create a nucleus of staff for such plantation and it will be needed to be trained in a plantation programme and in a maintenance programme for such tree plantation. For the first 3-5 years, the plantation programme will be self-sustaining and should possibly be revenue earning but not revenue spending. I would also recommend that a *Director of Horticulture Operations for Railway-sides Tree Planting Programme be appointed for the purpose.* If more details are required, I will be willing to have a discussion with the officers concerned, or the Forestry and/or Horticulture Department can give the required guidance.

8 Recommendations

These recommendations are not only for MPKV or for Maharashtra but for the whole of India. These are given in all humility but with forcefulness, and a request is made to all concerned to study them carefully, even though they may look familiar, and to give some urgent thought to their implementation.

Use Soil and Land According to their Capability

Every hectare, acre or even square metre of land is a resource and must be used according to its capability which will ensure its optimum and efficient use without any deterioration. The science and art of soil survey, soil and land evaluation is known, and is developing fast with modern technology.

No land is waste at the worst it is idle. During last 25 years (1982–2007), SPWD Society for waste land development has been working throughout the country and doing a commendable work in efficient wasteland development. It is idle because we are either lazy or helpless. It is my strong recommendation that all areas of land, whether agricultural range, forestry, water bodies or recreational, must be planned, conserved and taken care of. Land is our biggest asset. A country will develop or perish depending on her will to take care of her land resources.

At the request of the Chief Secretary's Committee on Land and Soil Resources in Maharashtra State, I was invited to give suggestions and I reproduce below the outline of these suggestions and recommendations.

Possible Objectives

1. Assessment and Evaluation of Land and Soil Resources of Maharashtra State.
2. Soil and Water Management for optimum production.
3. Soil and Water Conservation on all categories of land with special reference to lands with poor capability.

Steps in a Programme

- First step is preparation of land resource inventory of Maharashtra.
- Second step – take away marginal and sub-marginal lands from agriculture to sylvo-pastoral.
- Third step is to carry on Soil and Water Management on each of the above categories.

Agricultural Lands	-	Irrigated (Perennial)
	-	Irrigated (Supplemental)
	-	Rainfed
	-	Dry farmed
• Fourth Step	-	Management of problem soils
	-	Saline, alkaline and waterlogged soils
	-	Eroded soils

Urgent and Immediate Action Programme Needed

For Irrigated Areas

1. Drainage must accompany irrigation
2. Land levelling
3. Water management
4. Canal lining
5. Multiple and relay crop planning

For Supplemental Irrigated Areas

1. Land levelling
2. Water harvesting and water conservation
3. Multiple cropping
4. Ground water to supplement surface irrigation

Rainfed Areas >750 mm Annual Rainfall

1. Water harvesting and moisture conservation
2. Crop planning – Farming systems programme
3. Ground water development

Dry Farmed Areas <750 mm Annual Rainfall

1. Water harvesting and moisture conservation
2. Ground water development
3. Crop planning – Farming system programme

Saline, Alkaline and Waterlogged Areas Reclamation and Management

1. Drainage plans
2. Gypsum and other amendments
3. Organics
4. Crop planning

Intensive Action on Increasing Organics – Rural and Urban Composting

1. All municipalities *must* prepare compost by mechanical or manual operated compost plants
2. Bone-meal plants
3. Compost in rural areas:
   (a) with dairy farms
   (b) all crop waste
   (c) Glyricidia and Sesbania plantations

Marginal lands –Sylvo-Pastoral

1. Farm and social forestry – nurseries
2. Grassland nurseries
3. Pond construction

AFFORESTATION IN LARGE BLOCKS FOR WATERSHED MANAGEMENT

Strategy for Optimum Water Resource Development

I agree with the statement and proposals made by Capt. Dastur of Aviation and Industrial Projects as also that "We have more wealth in water than Arabia has in oil. We are a poor country because we have not bothered to look for and exploit our resources efficiently. And resources are there, very visible and yet mainly wasted."

A bold and essential plan has been proposed. It aims to stop floods and end droughts, both of which in turn and sometimes in quick succession plague the country. This plan or another more suitable plan or a combination of plans can employ the idle millions and other underemployed millions and settle even up to 300 million in lands reclaimed from the vast deserts and marginal and even poor lands irrigated and made fertile. The projects could provide cheap and literally limitless power throughout the country for industrial and domestic purposes.

It is recommended that while the bold projects given by Capt. Dastur or Dr. K.L. Rao on the Ganges-Cauvery Link, or others are discussed, finalized and blue prints for action programmes prepared, a survey of vast surface and ground-water resources must be completed with urgency even through satellite imagery. Large, medium and small reservoirs must be built in all parts. Small reservoirs in millions can be built as farm ponds, *nallah* bunds, ponds-*cum*-wells and other water harvesting means. Groundwater must be utilized fully as at present only a small percentage is being utilized (treat and over progress in last 30 years).

The efficient use of water through canal lining, water use

according to crop water requirement, through sprinkler, drip or the *ooze irrigation* method should be encouraged where applicable.

Develop and Nourish – MINRA

The Maharashtra Institute of Natural Resources for Agriculture (MINRA) could be an appropriate institution to serve and service Maharashtra's agriculture, specially as regards the following five focii of agricultural development :

(1) Soil testing, crop and farm planning and yield programming.
(2) Environmental control and recycling of wastes.
(3) Natural resources survey through satellite imagery.
(4) Social and farm forestry and range development.
(5) Training, consultation and extension in Natural Recources survey, development and utilization.

It is recommended that a number of devoted and capable personnel be recruited for this Institute, but more than that, it is recommended that the Institute becomes the embodiment of cooperative work between all the four Agricultural Universities, the Department of Agriculture of the State Government, local authorities, all determined to serve the small and even large farmers in the field of natural resources development for agriculture in Maharashtra.

It is also recommended that the Institute should use both modern and intermediate technology, be an innovator, coordinator and a training ground for optimum resource development. I am reminded of the delays that have slowed down development due to the delay in decisions and possible opposition from those unwilling to change. It is strongly recommended that to delay the establishment of an Institute like this would be detrimental and costly and against the interest of agricultural development in the State.

Further it is recommended that a similar Institute be developed in the different with, as applicable, after similar studies or projects are undertaken.

Multiple Croppings the Best Industry of the State and the Country

The abundant solar energy resource, is available in Maharashtra, has the capability to meet energy needs substantially beyond the application now being made. This energy on an average day is equivalent to 6 to 8 kilowatt hours per square metre. For normal plant photosynthesis, we are making use of 1/1,000th of the energy that is available to us.

To make better use of this energy source, we need to have more crops per year and more yield of each crop, i.e. greater production per day.

Multiple cropping is possible in larger and larger areas. The irrigated areas of Maharashtra, unless occupied by a perennial crop or trees, can support three annual crops and sometimes more, if relay crop technology is efficiently utilized. For rainfed crops, two crops are possible in most cases.

It is strongly recommended that emphasis be given to:

1. Multiple cropping for agricultural areas of the State.
2. Yield programming by soil testing and appropriate technology through a package of practices and inputs.

With the increased agricultural production, the unemployment and underemployment of people can be reduced considerably. Various field operations, post-harvest technology, storage and marketing all require more manpower and man-days and thus better distribution of wealth in the form of production.

Youth Corps – Need for Fuller Utilization of the Energy of the Youth for National Development

While some emphasis is being given to the soil, water and other natural resource development, India is neglecting her greatest asset, human energy. The time of maximum energy and vigour is between the ages of 15 to 25. It has been estimated that over 120 million Youths* of both sexes fall into this age category. The utilization of this vast resource of energy for national development is at a low level of efficiency and this must be improved.

It is recommended that steps should be initiated towards developing a "National Youth Corps of India"** whereby youth between the ages of say 15 to 25, all spend a two-year period for national service. It is proposed to have four Youth Corps :

1. Education Corps;
2. Health and Family Planning Corps;
3. Agricultural Development Corps; and
4. Rural and Urban Construction Corps.

Some ambitious projects such as the Great Himalayan Canal or the Ganges-Cauvery link or millions of pond construction could be part of construction projects. Planting of millions and even billions of trees and making them survive, compost pits preparations, control of pests and diseases etc. could form part of the task of the agricultural corps.

Preventive and even curative health plus family planning would be the task of the Health and Family Planning Corps. The speed and success of family planning activities will determine the course of events of national development and even the survival of the existing type of government.

A school in each village and imparting through such schools elementary and also adult education could form part of the terms of reference of the Education Corps.

---

* Now much more (2007).

** Even after 30 years of this recommendetion, the (NYCI) is even more valid and needed for the country's development. See a note on IAVE-India Association for Voluntary Effort of which I am a trustee. Note's prepared for the present situtation of Volutary Effort and some plans for International Centre for Voluntary Effort.

Such corps have been established in some other countries, need military disciplined training say of 6 months and work in villages for 18 months.

It is recommended that Pilot Projects be started for a Youth Corps programme in one or two States and even those progressively, before the project is started on a large scale and its financing be given greater thought. It is recommended that the ideas expressed in the paper on Youth Corps a Blue Print for India (presented in Appendix II) and associated papers be given greater consideration.

Working Hours in our Working Places need Radical Change

I have worked in various parts of the world and studied the relationship between the efficiency of work and working hours.

I would like to make the following observations:

1. Working efficiency is related to the temperature—both air-temperature outside and the temperature of the room. It is also dependent on humidity and radiant temperature and should be termed as effective temperature which is measured and is dependent on temperature, humidity and radiant heat.
2. When the temperature rises beyond a certain degree, efficiency goes down considerably.
3. 82°F is optimum for physical work efficiency.
4. 73-74°F is optimum for brain work efficiency and that is why air-conditioning is needed for optimum efficiency in many parts of India.
5. Measurement of physical efficiency is possible and certain tests have been designed. Mental efficiency is very difficult to be measured as there are no objective tests.
6. The question of rest or break between hours of work is important.
7. The normal daily temperature varies from city to city. The offices of the Government of India and the State

Governments are concentrated in New Delhi and the State capitals where an estimated 60-70 per cent government personnel work. The rest are in district headquarters, taluka headquarters etc.

8. If, in a country like Germany, the working hours can start at 7 to 8 a.m. even in difficult temperatures like in winter, why can we not open our offices earlier as proposed.
9. The efficiency of the operation of government officials at varying levels has a great influence on the *speed of development* in the country. Such efficiency is governed by :

   (a) Terms of reference of each official which should be clear, i.e. duties and responsibilities as well as authority.
   (b) Devotion of individual official – for which leaders or senior officials have to give a lead by personal example.
   (c) Working hours.
   (d) Working conditions.
   (e) Emoluments.
   (f) Incentives.

10. It is proposed that the total 40 hours work per week be timed as follows:

Large Cities	Summer	7-30 a.m.	to 2-30 p.m.
	Winter	8-30 a.m.	to 3-30 p.m. or 9.00 am to 4.00 pm
Small Cities	Summer	7-00 a.m.	to 2-00 p.m.
	Winter	8-00 a.m.	to 3-00 p.m.
Work on close to site		8-00 a.m.	to 12-00 noon
		14.00	to 17-00 p.m.
Work related to Rural Area		7-00	to 12-00 noon or 7.00 to 11.00 a.m.
		15-00	to 17-00 p.m. or 16.00 to 19.00 p.m.
		17.00	to 20.00 p.m.

The efficiency between 7-00 to 11-00 a.m. is much greater (about 2-3 times) than between 11 a.m. to 2 p.m. or 3-00 p.m. in most parts of India during 8-9 months of the year.

11. In these days of electricity shortage, the greater the concentration of working hours during day light hours, the greater the economy of electricity. For large cities, air-conditioning is a huge consumer of electricity. In most places during the first few hours of the day, there may not be any need for air-conditioning, fans etc.
12. Through such working hours, evenings are free for relaxation, sports, shopping and other cultural activities. These are essential for a healthy outlook on life and work.
13. The argument for distances from place of work is not tenable as public transport can be geared to this purpose. If private enterprise or schools can start earlier, or factory shifts, why not the offices of the government. Arrangements for food can be made as need of the timing dictates. Indians in general are early risers, excepting some big city dwellers and this is an unhealthy trend (late rising, bed tea system etc.).
14. In terms of priority, the task which is most important should get the highest priority. The job for which several million government servants are paid by the government is of higher importance than relaxing in the most important (early) hours of the morning.
15. I dare say that India will not awake and make enough progress unless the working hours are advanced in the office by 2 or 3 hours without changing the total number of working hours per week. Government offices are where bureaucrats and decision-makers work and that is where the tone and trend of efficiency has to start for the country as a whole.
16. The fact that 1-5 per cent of officials work at home in the early hours does not seem very relevant, as office work should not intrude on domestic time.

17. Lazing in the morning, starting with bed tea, doing some shopping, attending to telephones, socializing and even having an early lunch are not conducive to a proper way of working. One cannot start a day with a full stomach and show efficiency of work.
18. It is my strong recommendation and an *earnest appeal* that the working hours be advanced considerably. This step is essential for the advancement of the country. Let the Government of India and the State Governments be bold and take an early decision on this point.

Give High Priority to Social Participation and Farm Forestry with Effective People for the Survival and Good Growth of Trees

In view of the fact the aridity is increasing in many parts of the country, specially drought prone areas, mainly due to indiscriminate cutting of trees and over-grazing of vegetation and that marginal and sub-marginal lands, not only in Maharashtra State but all over the country, are increasing and the land capability classification is in general going down. It is strongly recommended that a *bold programme of social and farm forestry be developed in all parts of the country*. Recent NGO (2006) called "Millions of Trees International" –"MOTI" be supported by all. In this connection, it is strongly recommended that a Department of Social and Farm Forestry is established in all Agricultural Universities in India. This subject should not only be taught as professional foresters are taught, but as a part of land use in association with agriculture and on marginal and sub-marginal lands in conjunction with agricultural lands.

A bold programme of tree planting is recommended whereby in Maharashtra alone 100 million trees are planted in the next 5-10 years. It needs an army of devoted people not only to plant but to protect and take care of the trees planted. The Agricultural

Universities can help in the training of personnel at all levels for such a vital programme.

The alternative is aridization of larger areas in the country and the floods-drought-floods cycle will continue to occur with greater and severer frequency.

It is also recommended that the technology be developed in association with the tree planting programme of how to conserve and get good production of fuel, fodder and horticultural and other economic benefits from trees adaptable to different regions of the country.

One practical way to promote social and farm forestry is to use a part or the whole of taluka seed farms as social forestry nurseries. This idea as indicated by Vice-Chancellor, Dr. M.S. Pawar should be pursued.

APPENDIX-I

Reprinted from the Proceedings of FAI/FAO Seminar held at New Delhi on 13/14 December, 1974

Composting of Urban and Rural Wastes

M.L. Dewan and *J.H. Dongale**

ABSTRACT

The estimated quantity of compostable waste material available in the country is about 2,200 million tonnes which can contribute about 22 million tonnes of plant nutrients. Although, there is tremendous potential, the present status of exploitation of these resources is rather poor, i.e. about 5 million tonnes of urban and 150 million tonnes of rural compost (less than 6 per cent of the total potential). In the process of optimizing agricultural production under limited availability of fertilisers, composts can play an important role, specially in combination with chemical fertilizers.

The difficulties involved in the utilization of human, cattle and farm wastes are enormous, mainly because (a) the human waste in rural India is seldom collected in an organized way, (b) two-thirds

---

* Mahatma Phule Krishi Vidyapeeth, Rahuri (Dist. Ahmednagar).

of the cattle dung is burnt as fuel which is in short supply, and (c) the same is true of a good portion of the farm waste, though a part of it is fed to cattle as roughage. Programmes of large-scale forestry and growing of fuel trees have to be strengthened several-fold for which there are immense possibilities. The same is true in the matter of collecting farm and crop wastes and composting them in an organized scientific way. Such efforts will also prevent environmental pollution caused by the discarding or burning of such waste material. Strong measures are needed for extending the preparation and use of compost. Cities, towns and large villages should organize efforts for collecting and processing solid and liquid wastes. Several measures are suggested in the paper for greater exploitation of these nutrient-rich materials.

The word 'compost' is from Latin word *compositum*, meaning a mixture or compound of various manuring substances for fertilizing land. The word is now taken to mean an organic compound made principally of farmyard manure and green leaf material, straws etc., processed by decomposing the mixture in specially made pits or heaps. The pits or heaps are prepared with these materials in alternate layers, using chemical fertilizers as a starter and catalyst to accelerate the decomposition. But now it has been found that a multitude of materials that are normally being incinerated as useless waste can be used of this valuable material, but also preventing environmental pollution caused by these discarding or incineration.

Qualitative Potential

There are innumerable sources of compostable waste materials. They can be classified as follows :

1. *Waste material from human habitation*: Night–soil, urine, sewage, sludge, sullage, and town refuse.
2. *Cattle and slaughter house wastes*: Cowdung, cattle urine, blood, meat-meal and bone-meal.
3. *Farm and rural waste material*: Farm waste, wheat straw,

millet stalks, cotton stalks, horticultural waste, village refuse, and waste from *gur* production.

4. *Agro-industrial waste*: Bagasse, saw dust, tobacco seed cake, rice husk, coffee husk, tea waste, waste from cotton products, press mud and waste from fruit and vegetable processing plants.
5. *Forest and horticultural trees litter*: Private forests, Community forests and Reserve forests (tropical, moist-tropical, sub-tropical and alpine).
6. *Others and special sources*: Sea weeds, water hyacinth, fish meal, marine waste, and special sources such as from plants like *Glyricidia maculata, Sesoania aegyptica, S. grandiflora,* etc. specially grown for the purpose.

It can be seen how comprehensive the list of waste materials is and how diverse these could be.

Quantitative potential – Some Indication of Available Quantities

The sources are very diverse; however, some indications about the vastness of the quantities that are likely to be available of important materials have been worked out. These are as follows:

Urban and rural human waste: Large quantities of human excreta will be available for compost making and through sewage systems, if steps can be taken to collect them. For example, Table 1 could form the basis for working out the total available quantity as it gives the per capita availability of this waste material per day.

Table 1: Per capita availability of human waste per day

Particulars	*Faeces (g)*	*Urine (g)*
Quantity in natural condition	133.00	1200.00
Quantity in dry condition	30.30	64.00
N	2.10	12.10
P_2O_5	1.64	1.80
K_2O	0.73	2.22

From Table 1, it is possible to work out the total potential of these waste materials, by correlating them to the total population of this country estimated in 1981 at 684.7 million. The total quantity of waste material and plant nutrients in terms of N, P_2O_5 and K_2O that could be made available from the whole country is estimated in Table 2.

Table 2 : Total human waste and plant food materials that can be obtained from the poulation of the whole country

(Million tonnes/annum)*

	Waste material		*Plant nutrients*		
	Dry	*Wet*	*N*	P_2O_3	K_2O
Human faeces	27.6	6.1	0.429	0.335	0.149
Human urine	245.3	14.7	2.473	0.367	0.454
Total	272.9	20.8	2.902	0.702	0.503
Say			2.90	0.70	0.59

* In 1981, 20 per cent increase in population over 1974 and hence 20 per cent increased potential and still 20-50 per cent incharged potential in 2007.

It may be seen that the total quantity of only major plant nutrients, viz., N, P_2O_5 and K_2O available is about 4.1 million tonnes apart from minor elements. At present, the materials discharged through underground sewage systems are used in sewage farms in some cities. There are 600 towns which have open drains. The sewage and sullage available in these cities is about 3,200 million litres per day, equivalent to a flow from 1,400 tube-wells which can irrigate about 1,00,000 hectares every year. Although sewage is being dealt with in a separate paper, it may be stated here that on an average, sewage contains 25 ppm N, 4 ppm P_2O_5 and 15 ppm K_2O and 400 ppm organic matter. The plant nutrients, viz., N, P_2O_5 and K_2O discharged annually through 3,200 million litres of sullage per day are about 0.3, 0.06 and 0.18 million tonnes respectively or total of 5,40,000 tonnes of plant nutrients. While the utilization of these resources in the urban areas is very limited, in villages, it is negligible. The village surroundings are at present defiled by human excreta creating insanitary condition and constituting a danger to public health. There are practically no conservancy arrangements in the rural area and the majority of the public is in general not using any type of latrines.

Cattle waste including city dairies and slaughter house wastes etc. The total cattle population of India is about 350 million, producing annually about 1,335 million tonnes of wet cattle dung. Cattle urine produced is about 370 m. tonnes per annum. The quantities of plant nutrients available in the form of cattle dung and cattle urine in India and quantities actually used are presented in Table 3.

Thus only about 20 per cent of the total plant nutrients are estimated to be used.

Table 3 : Quantity of plant nutrients available and used in the form of cattle dung and urine

	Potential			*Actually used*			*%*
	N	*P_2O_5*	*K_2O*	*N*	*P_2O_5*	*K_2O*	*Used*
	(million tonnes)			*(million tonnes)*			*estimated*
Cattle Dung (0.2% N, 0.1% P_2O_5, 0.15% K_2O)	2.6	1.33	2.0	0.89	0.44	0.66	33
Cattle Urine (0.6% N, 0.01% P_2O_5; 0.5% K_2O)	2.22	0.04	1.85	0.04	0.02	0.03	5
Total	4.89	1.37	3.85	0.93	0.46	0.69	
$N+P_2O_5+K_2O$			10.11			2.08	

Nearly 40 million animals are slaughered and about 12 million carcasses of large animals are available annually. In addition, about 55,000 tonnes of blood is available annually, about two-thirds of which is generally wasted. Approximately, 1,20,000 tonnes of meat meal can be produced from animal carcasses. Bones which constitute approximately 15 per cent of the weight of carcass, are used for preparing bone-meal. The total potential for animal bones is about 4,50,000 tonnes annually of which less than 30 per cent or 1,36000 tonnes is collected. Five thousand tonnes of leather waste is also available which remains uncollected. The estimated available quantities of these wastes as also of fish meal, poultry droppings, sheep and goat manure, etc., are presented in Table 4.

Table 4: Slaughter house waste and animal and fish waste products potential (per annum)

	No. (million)	*Total weight (tonnes)*	*Composition Per cent*			*Production potential (million tonnes)*		
			N	*P_2O_5*	*K_2O*	*N*	*P_2O_5*	*K_2O*
Slaughtered Animals40								
Natural death	12	2.4 (million)						
Blood	-	55,000	10-12	1.2	0.5-1	0.006	0.006	0.006
Meat meal	-	12,000	10.15	2.5	0.5	0.0025	0.005	-
Hoof & Horn	-	12,000	10-15	1	-	0.0025		
Bones	-	45,000	3.5-4.5	18-25	-	0.02	0.100	-
Leather waste	-	5,000	7.00	0.1	0.2	0.003		
Hair & wool waste	-	N.A.	12.3	0.1	0.3			
Fish waste	-	40,000	4-10	3-9	0.3-1.5	0.003	0.02	-
Poultry droppings	114	40,00,000	7.5	6.5	3.5	0.3	0.26	0.14
Sheep manure	40	30 (million solid)	0.70	0.51	0.51	0.21	0.15	0.09
Goat manure	60	15 (million liquid)	1.47	0.05	1.96	0.22	0.08	0.30
Total						0.76	0.62	0.54

Farm Crop Wastes

These usually include sugarcane trash, stubble of jowar and other crops, weeds, spoilt fodder, trash horticultural gardens, etc., Often, wheat, paddy and other straws which are spoilt and not fed to the cattle are included in the farm crop waste.

Some individual crop wastes have been assessed as regards the total quantity available but no general figures are available. A study was made at the central farm of the Mahatma Phule Krishi Vidyapeeth, Rahuri and at Agricultural College, Kolhapur on the quantity of crop waste available and which could be used for composting. The data are presented in Tables 5 and 6.

Table 5 : MPKV Central Farm Crop Waste

Crop (Both Kharif+Rabi)	*Area (Hectares)*	*Estimated expected waste material in tonnes (Stalks+Straws)*	*Average waste (tonnes/ha)*
Sorghum (jowar)	179	1700	9
Bajra	53	271	5
Sunflower	236	726	2
Sunhemp	76	153	2
Sugarcane	44	225	5
Groundnut	121	392	3
Wheat	284	2620	3
Safflower	41	92	2
Gram + Pulses	250	367	1.5
Weeds	—	310	—
Total cropped area	1318	6856	

Table 5A: Kolhapur Farm Crop Waste (Waste in Tonnes)

	Agrils College Kolhapur (Area 34 ha.)	*Shende Park (40 ha)*	*Regnl.S'cane Res. Station (25 ha)*	*Total (99 ha)*
Sugarcane Waste	1.5	38	15	54.5
Rice Straw	5	12	7	24
Other Crop Waste (Stubble, vegetable stalks)	4	1	4	9
Weed	7	7	10	24
Total	17.5	58	36	111.5

Waste material can be used partly in the industry such as rice, straw, sugarcane waste for paper, etc. Some wastes which are being used as fuel can be discouraged. On an average, the quantity of crop waste available per hectare is about 1.1 tonnes although this can go upto 9 tonnes per hectare depending on the crop, i.e. whether it is sugarcane, or sorghum or wheat etc. Even if we assume an average of 1 tonne of waste material per hectare, taking into account the area under important crops such as rice, wheat, maize, sugarcane, cotton, potato, sorghum and pearl millet, which total 101.4 million hectares, the total production of crop waste would be as follows:

87.3	million tonnes from cereals,
11.73	million tonnes from sugarcane,
1.17	million tonnes from cotton,
4.83	million tonnes from potatoes.

Total plant nutrients available from these waste materials

	N	P_2O_5	K_2O
Million tonnes	3.6	2.3	10.7

Table 6 : Agro-Industrial wastes

Item	*Quantity (million tonnes)*	*Plant food content (%)*		
		N	*P_2O_5*	*K_2O*
Oil cakes	2	3-6	1-3	1-2
Wastes from fruit and vegetable processing industries (peels, stones, skin, shell etc.)	N.A.	N.A.	N.A.	Variable
Bagasse	13.5	0.25	0.12	—
Press mud (from sugar factories)	2	1-1.5	4-5	2-7
Saw-dust manure	N.A.	N.A.		
Tobacco waste	.062	0.5-1.0	0.8	0.8
Rice bran	2	N.A.		
Tea wastes	N.A.	2.8-3.6	0.3-0.4	1-1.9
Coffee wastes	N.A.	N.A.		
Cotton dust and textile waste	N.A	1-1.5	—	—

N.A. = Not available

It is, therefore, clear that great potentiality exists for recycling crop wastes.

Industrial Waste

Especially Agro-based Industries

These include industrial wastes especially from Agro-based Industries and are given in Table 6.

Forest and Other Horticultural Trees Litter

No accurate data is available on forest litter of various types as well as from fruit trees such as mangoes etc., which shed dry leaves every year. It is estimated that about 13 million tonnes of compost can be obtained from the forest litter annually. They may contain 75,000 tonnes of N, 30,000 tonnes of P_2O_5 and 75,000 tonnes of K_2O. Most of the litter is burnt and thus huge quantities of plant nutrient go waste.

Other special sources include sea weeds, water hyacinth, marine waste, also special sources such as *Glyricidia* or *Sesbania* have immense potentialities.

No proper assessment to the extent of availability has been made so far. Available figures are presented below:

	Present quantity collected	*Potential*	*Plant food content (per cent)*		
			N	*P_2O_5*	*K_2O*
Sea weeds	15,000 tonnes	N.A.	—	1-2	2-1
Water Hyacinth	N.A.	N.A.	N.A.	N.A.	
Marine waste	N.A.	Considerable	0.3	0.3	0.3
Glyricidia	N.A.	Immense	0.53	—	—
Sesbania	N.A.	Immense	0.40	—	—

N.A. = Not available

Summary of Potential

The potential for exploitation of urban and rural wastes may be summarised in Table 7.

Table 7 : Urban and Rural Wastes

	Quantity (Million Tonnes)		N	P_2O_5	K_2O	Total Plant
	Wet	Dry				
1. Human waste (faeces+ urine)	273	21	2.90	0.70	0.50	4.1
2. Cattle (dung+urine)	1,705	150	4.89	1.37	3.85	10.11
3. Slaughter house wastes,						
Carcass by-products	4		0.20	0.30	0.10	0.6
Fish meal	1		0.06	0.05	0.01	0.12
Poultry waste	4		0.3	0.28	1.4	1.98
Sheep and goat manure	100		0.6	0.20	1.6	2.40
4. Crop waste	100*		0.6	0.20	1.0	1.8*
5. Agro-industrial waste (estimated)	20		0.2	0.10	0.2	0.5
6. Forest and horticultural trees litter	15		0.05	0.04	0.08	0.2
7. Other sources	10**		0.1	0.1	0.2	0.4**
Total	2232	435	9.93	3.57	7.34	22.14

* Can be much more with higher yields.

** Can go upto 100 instead of 10.

The data given in Table 7, though not complete, show the immense potential that exists for utilization of these wastes.

Project Plans

The use of chemical fertilisers has steadily grown in India from about 1,50,000 tonnes in 1950-51 to 3 million tonnes in 1974 and is estimated to go up to 8 million tonnes by 1979-80, out of which 6.5 million tonnes is expected to be produced in India, Optimum use of fertiliser is essential, but production of composts and organic presents a suitable additional source for making up the deficiency of plant foods. Because of the size and its population, the country has enough waste material like cowdung, night soil, urine, forest litter, sewage, industrial and organic wastes, city garbage and crop residues.

The present and project plans for production of urban and rural compost are given below:

	1974	1979	N	P_2O_5	K_2O
	(million tones)			(Per cent)	
Urban Compost	4.45	75.0	1.5	1	1.5
Rural Compost	158	350	1.5	1	1.5
Government Budget for Composting					
(million rupees)	90	180	_	_	_

Prospects and Current Status

The prospects are immense as discussed earlier. Current status indicates that composting is becoming increasingly popular, but it is still not enough. It is only a small percentage of waste material collection, composting and the time lag involved act as deterrants in the speedier adoption of composting. However, shortage of chemical fertilizers ensure good crop and better yields. Composts increase the nutrient resources of the soil and serve as sources of micro-nutrients. They also play a vital role in improving the soil structure, water holding capacity and resistance to soil erosion.

Three hundred and twenty large compost pits were, for example, made in one Agricultural University campus at MPKV, Rahuri in one month. A large number of extension workers and students were trained in scientific composting which is helping the region in increased compost preparation. Such efforts must be multiplied a thousand time so that compost preparation and utilization will have a significant impact on crop production even without use of chemical fertilizers or possibly in combination with them.

Technology Available for the Compost Preparation

The principles underlying the production of fermented bulky organic manures from cattle shed waste have been extended to the preparation of compost in which a variety of rural as well as urban and industrial waste materials may be utilized.

Thus, farm wastes like straw, chaff, leaves, paddy and groundnut straws, sugarcane straw, bagasse and other agricultural as well as rural and human habitation wastes like night-soil, slaughter house waste, sewage and sludge water hyacinth, etc., can be put into the compost pit. In addition, certain quick growing trees and bushes and certain weeds which are commonly found in rural areas on the fields can be composted. The technology for this may be described under the following methods:

1. Indore method of composting takes about three months for preparation.
2. Compost from habitation waste – Bangalore process – takes three to six months period.
3. Preparation of synthetic compost – ADCO process.
4. Mechanical composting.
5. Super digested compost.
6. Enriched organic based compost and their preparation.
7. Use of manure – biogas preparation.
8. Special process for use of fish meal, leather waste, hair and wool waste etc.
9. Use of *Glyricidia* and *Sesbania* for compost.

The procedure and methodology are to be covered elsewhere.

Problems of Organisation

These include: (i) Problems of research and development, (ii) Problems of extension under (i) are included:

(a) Development of efficient, integrated nutrient supply systems under different agro-ecological systems based on an appropriate mixture of chemical, biological and organic sources of nutrition and appropriate choice of crops and varieties.

(b) Development of quick yielding fuel trees in order to eliminate the need for using potential organic manure as fuel.

(c) Development of material for an expanded utilization of the process of biological nitrogen fixation.

The rate of crop removal (25-30 kg N, 11-12 P_2O_5 and 37-68 K_2O per hectare) can be theoretically met by recycling human and animal habitation wastes on agricultural land as this recycling can give 42, 11 and 25 kg/ha of N, P_2O_5 and K_2O, respectively. However, there are enormous difficulties in collecting all the potentially available cattle dung and urine as well as human faeces and urine. The availability of nitrogen added in the form of cattle manure to the current crop is only about 3.3 per cent of its N content, about 66 per cent of its P_2O_5 content and 75 per cent of its K_2O content.

Another important problem is to *accelerate the mineralisation of organic manure in the soil to meet the demands of growing crops* and at the same time preserve its favourable effects on the physical condition of the soil.

The difficulties involved in the utilization of human and cattle wastes are enormous mainly because (i) the human waste in rural India is seldom collected in an organized way as in developed countries, (ii) two-thirds of the cattle dung is burnt as fuel is in short supply.

For the first, socio-economic and educational conditions will need to be reorganized before human waste in rural India can be collected and recycled.

For the second, large-scale farm forestry and growing fuel trees are essential for utilizing cattle dung for composting and recycling back to the soil.

Promotion of biogas plant both small and large has to be stimulated whereby both animal and human wastes can be utilized. Fuel (methane) for home cooking is produced and effluent is readily available organic fertilizer.

Recommendations

(a) It is recommended that each taluka or at least each district should have a *compost unit* which will stimulate and do extension work for the increased collection of waste materials and composting them and standardizing compost pits which should be enriched by special process developed. Large scale audio-visual methods and demonstrations should be conducted for this purpose.

(b) It is recommended that cities having over 1 lakh population should have mechanical compost plants. Compost from these plants, which is enriched should be available for sale to vegetable growers and/or cash crop growers near the city.

(c) It is recommended that sewage plants be installed in all cities and towns having more than a population of 25,000 and sewage material be used efficiently.

(d) A law may be passed to make essential the composting of all wastes and secondly, to make it punishable if a waste of more than a certain quantity is found in a particular farm or village unit.

(e) Larger quantities of *Glyricidia, Sesbania and other bushes and crops be grown especially in waste lands, bunds etc. to produce large material for composting.*

(f) It is recognized that bulky organic manures, in particular play an important role in increasing crop yield through (i) supply of plant nutrients, and (ii) effecting improvement in several aspects of soil fertility like the physical, chemical and biological properties of soils. Variable and often non-significant responses that are obtained through the use of these composts and manures are now understood to be mainly due to the low availability of nitrogen in them and consequently their inability to make good the primary deficiency of nitrogen in the quantities applied. *It is, therefore, recommended* that effective use of such composts should be made by: (i) use in adequate quantities, if available, on the basis of their plant utilizable nitrogen content, (ii) use in combination with fertilizer nitrogen in quantity to provide adequate nitrogen for the requirement of crops, and (iii) use after enrichment with nitrogen and phosphates as described earlier. Since the quantity of manure available with farmers is considerably short of their requirements, the latter two alternatives provide a solution that would give an assured response.

SUGGESTED READINGS

Bawaskar, V. S. and Zende, G.K., The effect of application of different dungs and waste material on crop yield, *Agriculture and Agro-Industries Journal.* January, 1971.

Bawaskar, V.S. and Zende, G.K., Soil fertility under continuous manuring and cropping, *Indian J. Agric. Sci.*, 43 (5), 492-499 (May, 1973).

Barooah, S.R., Organic manures : A substitute for fertilizers, *Agriculture and Agro-Industries Journal,* 33-34 (April 1974).

Biswas, T.D., Cowdung gas plant for energy and manure, *Fertilizer News,* 19 (9), 3-7 (September, 1974).

Garg, A.C.; Idnani, M.A. and Abraham, T.P., Organic manures, ICAR, *Technical Bulletin* (AGRIL). 32 (April, 1971).

Indian Council of Agril. Research, *Handbook of Manures and Fertilizers,* ICAR, New Delhi (1971).

Panse, V.G. and Abraham, T.P., Green manuring of crops, ICAR. *Tech. Bulletin,* No. 2 (1965).

Swaminathan, M.S., Organic manures and integrated approach to plant nutrition, Consultative Group on International Agricultural Research Technical Advisory Committee (July-August, 1974).

China, Recycling of Organic Wastes in Agriculture—Report on an FAO/UNDP Study Tour to the People's Republic of China, 28 April-24 May 1977, *FAO Soils Bulletin No. 40,* Rome, FAO (1977).

China Azolla Propagation and Small-scale Biogas Technology – Report on FAO/UNDP Study Tour to the People's Republic of China, 21 May-11 June 1978, *FAO Soils Bulletin No.41,* Rome, FAO (1978).

APPENDIX-II

Youth Corps – A Blueprint for India

*M.L. Dewan** and *Vikram Dewan***

While the world speaks of an energy crisis, India is neglecting her largest asset: human energy. But how can this physical energy be properly utilized ?

The time of maximum human energy and vigour is between the ages 15 to 25. It has been estimated that over 120 million youths of both sexes fall into this age category. But it can be seen that this youthful force can be destructive witness, for example, university strikes – unless it is mobilized and institutionalized.

Such a possible institution is the Youth Corps programme. The purpose of this programme would be three-fold. Primarily, there exists a need to mobilize students and unemployed youth for development projects. Secondly, such a programme could foster participation in the improvement of the health and welfare of the Indian people. Finally, the Youth Corps could catalyze the advent of mass education.

To accomplish these aims, the Youth Corps would be divided into four major categories.

'The Literacy Corps' objective would be to assist in the mass education programme to achieve total literacy in India within a specified time period.

The Health and Family Planning Corps would help in spreading

* Soil Scientist, FAO of the United Nations and Visiting Professor, MPKV, Rahuri.

** International Development Centre, Cornell University Ithaca, New York, USA.

and understanding of the ways and means for population control, as a supplementary agency to the currently existing family planning centres. This Corps group would be directed towards introducing preventive and curative means in health services by providing for the rural, and the poorer sectors of the urban populations.

Similarly, *the Agriculture Production and Development Corps* would extend modern agricultural knowledge by initiating a large number of production centres. Without displacing farmers, this Corps group could be allotted reclaimed land under irrigation on which pilot projects could be undertaken. Afforestation and soil conservation measures would be effected by this group.

The Housing and Construction Corps would build dwellings and apartments for the rural and urban populations. Although local materials should be used, this Corps would assist in building rural roads to aid the construction process. In addition, major projects such as bridges, schools, family planning centres, wells and reservoirs could be built by this Corps group.

Some of the specific tasks that the Housing and Construction Corps could undertake could be a speedy completion of the Rajasthan Canal Project, or to initiate construction on the Ganges-Kaveri Link Project, or work on large and small water resource development projects, or build houses for the masses.

The Agriculture and Development Corps could commence reclamation of the Chambal Ravines and undertake large-scale afforestation. In assisting the Environmental Control Project, the Corps would recycle wastes for agricultural production purposes.

The Literacy Corps would aid the Housing and Construction Corps in building a school in every Indian village which does not already have a school.

To ensure the success of these various Corp groups, the youths would elect leaders from amongst themselves. These leaders would possess university degrees in their respective fields and the qualities of leadership that would elicit obedience from the participating youths.

The idea of mobilizing youth energy is not a new one. This notion has been attempted and implemented in many nations. The Youth Corps in Iran, started in 1964-65, contributed significantly to the Shah's White Revolution as evidenced by significant developments, including a rise in the standard of living of the common people. The rise in Iran's GNP is not due to oil alone although oil revenues have provided the financial base in developmental activities. The Land Army initiated by Chairman Mao, in China, is another example of this effort to mobilize youth for development. "Brigades" in Bulgaria are a similar example whereby students and youths are employed during the summer in construction projects as well as crop harvesting. This institution has proved to be a great boost to the Bulgarian economy.

But instead of following anyone of these paths, India can develop her own brand of Youth Corps. Our problems are neither unique nor unsurmountable and, given India's greatest source of energy, the Youth Corps offers a viable solution to our development scene.

"Much detailed work is possible in the utilization of idle capacities and the implementation of already adopted schemes" claims R. Kothari of the Centre for the Study of Developing Societies in Delhi. Dr. Kothari elaborates that "the education system offers unlimited scope for improvement, and the vast masses of unemployed and underemployed in the country can be turned into a powerful force for rapid development by initiating specially tailored programmes". The Youth Corps movement in India is such a programme.

The emergence of the National Cadet Corps (NCC) was a recognition of the need to mobilize the youth for constructive action. However, a broader and more concentrated programme for both sexes of youth at all educational levels is needed, and the National Cadet Corps can provide a case for building up such a Youth Corps with extended tasks and activities and be formed into one single National Youth Corps organization, in which other similar youth development activities could be amalgamated. They could in addition offer a second line of defence, also needed for our country. This would be possible when the army takes the major share of training – and this is what is proposed.

Mr. R.K. Karanjia, Editor of *Blitz*, asserts that India is wasting her precious natural resources – her people and particularly her youth. Citing China as an example, Mr. Karanjia indicates how Mao's Land Army became an example of mass participation in the development as well as implementation of land reforms, whereby agricultural production increased two or three-fold. Mr. Karanjia believes that it is a tragic paradox that, while vast and untapped natural resources require human effort, there exists unemployed, immobilized youth.

Dr. Karanjia pleads for the adoption of Gandhi's "Gospel of the Dirty Hand", whereby three million college and university level students along with 59 million other youths in the same age category could form a Land Army. These Land Armies could operate on the basis of a grand cooperative; participation could encourage the values of self-help, self-reliance and self-government. While university education budgets could assist in financing the project, the State would be the main source of revenue.

Adequate preparation is needed to ensure the success of the Youth Corps as a development institution. Initially, there could be a voluntary small-scale training programme which could later be expanded to include all students on a Government budget. In the pilot phase, 1,000 students could be trained for two weeks to work on designated projects for the remainder of that summer. Henceforth, members of the unemployed youth could undergo a one month training to assist in development projects for eight months. The training facilities and personnel could be provided by the universities involved in the programme.

For the three thousand youths involved in the pilot project, expenses would be approximately three million rupees per year. This sum would be secured through university and other private sources.

Ultimately, it is hoped that an organ of the Government will handle the activities of the Youth Corps. There should exist training centres in each State clarifying the tasks of this organization. Inculcated with sense of discipline, and independent of a government bureaucracy, the Youth Corps could emerge as a major development force before the end of the 1980s. The goal is a self-sustaining Youth Corps whereby the national benefit is more than the cost of labour and upkeep of the Youth Corps organization.

The necessity of the Youth Corps for India can be highlighted by the following sentences:

> "It is universally admitted that our democratic system and economic patterns have not yielded the required results because of the lack of human mobilization and mass participation behind costly plans and projects. In this context, massive land armies of the unemployed, particularly youth, provide God's own solution to the main problem slowing our development."

APPENDIX-III

Student Training and Orientation for Leadership in Accelerated Agricultural Development (Stolaad)

Vikram Dewan

Aims and Objectives

1. To provide a forum for leadership among the student groups for training and orientation in means and measures for speedy agricultural development in various parts of India.
2. To provide a long-term national perspective for student participation in efforts such as Dignity of Labour Camps, Land and Water Conservation and other development activities.
3. To create conditions favourable for having a multiplier effect on student participation.

Organization

Location

Each university, such as MPKV, could be the centre of such a training and orientation course for periods ranging from one week to a trimester.

Progressive Development

Such a training course for students leadership is to be developed in the following stages:

Stage I : Within the MPKV Central Campus at Rahuri and secondary schools and colleges in neighbouring areas.

Stage II : At agricultural and other universities and associated educational bodies throughout the State of Maharashtra.

Stage III : Same as stage No. II, but on an all India basis.

Selection of Participants

The participants are to be chosen on demonstrated ability with regard to interest, output of work and qualities of leadership. For the first stage, the nine schools, three associated colleges and the MPKV Central Campus, twenty research stations, ninety-six secondary and five colleges in Ahmednagar district will provide the bulk of participants. For the second stage, the Universities will nominate the participants. For stage three the States will nominate the participants.

First Step

It is proposed to start a one week orientation course for student leadership, for about twenty students at a time. It is expected that such a training course will be conducted initially to cover the MPKV area of operation.

Components of Agricultural Development

Soil and Water Conservation, Management and Development

1. Contour bunding
2. Graded bunding
3. Gully bunding

4. Nallah bunding
5. Farm ponds
6. Stone collection and putting along contour

Water Management

1. Canal realignment and lining
2. Land levelling
3. Drainage

Tree Planting

1. Planting of fuel, fodder and horticultural species
2. Large-scale afforestation
3. Maintenance of trees

Recycling of Waste

1. Compost preparation
2. Organic recycling including use of farm waste
3. Use of urban waste
4. Biogas plant installation

Livestock, Poultry and Inland Fisheries

1. Silo preparation
2. Use of livestock and poultry waste
3. Development of inland fisheries

Training and Orientation Course

Leadership Training

This will be provided by actual work in the field in on-going land

and water development and conservation activities at the Central Campus. All the participants will spend 50 per cent of their time on actual work and the other 50 per cent will be devoted to lectures, group discussions and presentations from each one of the participants. Such a presentation should include a measure of individual in order to develop and improve the course.

Student Power Utilization

To bring out in due course:

- the availability of student power all over the country;
- to quantify such energy;
- to give the means and measures for the efficient utilization of the student power.

Create Conditions for Producing the Multiplier Effect

This consists in conveying the urgency of efficient utilization of student power to the policy makers in the Government through varying levels, down to the farmers, and in specifying the problems and impediments involved and a discussion on how exactly to overcome them.

It is proposed that the lectures will be for leadership training, and that seminars will be the means to the understanding of a national perspective as well as to create a multiplier effect.

Selection of Personnel

This will be on the basis of recommendations of the constituent colleges/schools and will be finalized by a small committee for this training and orientation course – STOLAAD.

APPENDIX-IV

List of Species to be used in the Afforestation, Farm Forestry and Social Forestry Project in MPKV, Rahuri

Sl. No.	*Scientific Name*	*Local Name*	*Purpose*	*Remarks*
1	2	3	4	5
1.	*Acacia Arabica*	Babul	CB, FW	Seeds to be collected local
2.	*A. Catechu*	Khair	GP, FW	Silva Poona
3.	*A. Auricaliformis*	–	OL	"
4.	*A. Planifrons*	–	GP	CCF, Tamil Nadu
5.	*A. Tortitis*	Israili Babul	GP	CAZRI, Jodhpur
6.	*Agave Sisaliana*	Ghaipath	Live hedge (LH)	LC
7.	*Acgle Marmelos*	Bel	Temple premises	LC
8.	*Ailanthus excelsa*	Maharukh	AV, LF, GP	Silva, Poona
9.	*Albizzia amara*	Tugili	GP, FW	"
10.	*A. Lebbek*	Siris	Bush sowing (BS),GP	LC
11.	*A. Procera*	White Siris Kinhai	AV, LF	
12.	*Anoma squa mosa*	Sithaphal	B C	CCF, AP
13.	*Azadirahta indica*	Neem	(AP)	L C
14.	*Bamboosa tulda*	Bihari bamboo	CB, SB	CCF Bihar
15.	*Bauhinia variegata*	Kanchon	OF	Silva, Poona
16.	*Buchananic latifolia*	Haroli	BS, GP	"
17.	*Capparis dccidna*		On Golf tops	"
18.	*Cassia fistula*	Bahava	OF	CCF MH
19.	*Cassia auricuiata*	Tharwad	(BC)	L C
20.	*C. Javanica*	Amalthas	OF	CCF, MH
21.	*C. Siamea*	Kashid	FW, BS, AV	Local C.
22.	*Casuarina equisetifolia*	Suru	FS, GP-OL, AV	Silva, Poona
23.	*Cochlospormum gossypium*	Canheri	OF	"

(*Contd.*)

(Contd.)

1	2	3	4	5
24.	*Calatropis prosera*	Rui	BC	L C
25.	*Chloroxylon sweitenia*	Satinwood Bhirra	BD	"
26.	*Cloistanthus collinus*	Garari	FW, GP	L C
27.	*Dalbergia sissoo*	Sisoo	AP	L C
28.	*Delonix rigia*	Gulmohor	AV, OF	"
29.	*Dendrocalamus strictus*	Kalki bamboo	CB, SB	"
30.	*Erythrina lithosperma*	Paugara	OF	E
31.	*Eucalyptus hybrid*	–	SB, SB, AV	-Do-
32.	*Emblica officinalis*	Avla	G P	Silva, Poona
33.	*Ficus bengalensis*	Kad	SP	L C
34.	*F. retusa*	Nandurk	SP	L C
35.	*Gmelina arborca*	Shivan	GP, FW	Silva, Poona
36.	*Grevillea robusta*	Silver Oak	OL & OF, AV	"
37.	*Glyricidia maculata*	–	LH, FW	Local
38.	*Hardwickia binata*	Anjan	FW, LF, GP	Silva, Poona
39.	*Holoptetia integrifolia*	Wawala	CB, SB, GP	CCF, Rajasthan, Jaipur
40.	*Ingadulca*	Vilayathi chinch	SP, LF, LH	LC
41.	*Jacaranda mimosifolia*	Jacaranda	OF, AV	Silva, Poona
42.	*Kigelbia pinnata*	Kigettia Sausage tree	OF, AV	"
43.	*Lawsonia alba*	Mehandi	LH	
44.	*Lencena glauca*		LF, GP	"
45.	*Madhuca latifolia*	Moha	GP, AV	"
46.	*Maugifera indica*		AV, CB, SB	"
47.	*Millingtonia hortensis*	Indian corkwood Akash Neem	AV, OF	Silva, Poona
48.	*Michelia chanipaka*	Chapa, Sonchapha	OF, Residential compound	" "
49.	*Morus alba*	Tulthi, Mulberry	CB, SB	"
50.	*Mimosaph Hexandra*	Rayani	GP, BS.	"
51.	*Ougenia dalbergiodes*	Tiwas	GP, LF	"
52.	*Polya lthia longifolia*	Ashok	OL, AV	"
53.	*Outgecokobium samah*	Rain tree	OF, AV	"
54.	*Plumaria acatifolia*	Pandra Chapha	Temple Planting	Local
55.	*Peltaphorum ferrugini*	–	OF, AV	Silva, Poona
56.	*Pongamia pinnata*	Karajj	CB, SB, AV	"
57.	*Prosopis cineraria (P. Spicigera)*	–	GP, LF	"
58.	*P. Juliflora*	Vedi babul	LH	Local
59.	*Santalum Album*	Chandan	BS	Silva, Poona
60.	*Sizygium jambolanum,*	Jambul	AV, CB	Local
61.	*Salix tetrasperma*	–	SB (in undulated areas)	Silva, Poona
62.	*Sesbania aegyptica*	–	LH, Windbreak, green manure	Local
63.	*S. Grandiflora*		"	"
64.	*Saraca indica*	Ashok	OF	Silva, Poona
65.	*Tamarindus indica*	Chinch	AV, GP	"

(Contd.)

(Contd.)

1	2	3	4	5
66.	Tecomella undulata	Rohida	OF, AV, GP	CF, Rajasthan, Jaipur
67.	*Tectona grandis*	Sagwan	GP	"
68.	*Thespesia populnea*	Bhendi	AV, CB	"
69.	*Thevetia nerifolia*	Bitti	LH	Local
70.	*Vitex negundu*	Nergudi	LH, Gully plugging	"
71.	*Zigyphus Jajisba*	–	Lca-cultivation	"
72.	*Butee monosperma (B. Crandosa)*	–	"	Silva, Poona

Note : CB=Canal bank, FW=Farmwood, GP=General purpose, OL=Ornamental, LH=Livehedge, AV=Avenue, LF=Leaf fodder, BS=Bushshwing, SB=Stream banks, OF=Ornamental flowers, BC=Broadcasting, AP=All purpose

APPENDIX-V

Operational Research Project on the Development and Use of Suitable Earthen Pitchers for Life Saving "Ooze Irrigation" for Newly Planted Trees

M.L. Dewan, A.V. Patil and *T.D. Patil*

Introduction

A Large number of trees must be planted in our marginal and sub-marginal lands in all parts of the country specially DPAP areas. Useful species of trees both for fuel and fodder and fibre should be as follows:

Sl. No.	*Scientific Name*	*Local Name*
1.	*Azadirachta indica*	Neem
2.	*Dalbergia sissoo*	Sisoo
3.	*Dendrocalamus strictus*	Bamboo
4.	*Acacia arabica*	Babul
5.	*Albizzia libbek*	Siras
6.	*Tectona grandis*	Sagwan
7.	*Agave sisaliana*	Sisal
8.	*Acacia catechu*	Khair
9.	*Pongamia pinnata*	Karanj
10.	*Hardwideia binata*	Anjan
11.	*Santalum album*	Chandan
12.	*Cassia siamea*	Kashid
13.	*Tamarindus indica*	Chinch

(Contd.)

(Contd.)

Sl. No.	*Scientific Name*	*Local Name*
14.	*Casurina equisetifolia*	Suru
15.	*Kangifera indica*	Mango
16.	*Punica granatum*	Pomegranate
17.	*Grewia asiatica*	Phalsa
18.	*Ficus carica*	Fig
19.	*Zizyphus mauritiana*	Ber

Above, list also includes some useful horticultural species.

Forestry Departments in various States have been working on such afforestation schemes from time to time but they are for the forestry areas and not in agricultural areas. The farm forestry relates to the planting of trees especially for farm areas which are part of the farmers' lands but which have become so much eroded that it is not profitable to grow field crops in these soils.

In the planting of trees there are a large number of tasks that are to be carried out.

Seedlings

This requires nursery operation and preparing seedlings which are healthy and withstand rigorous conditions. Nursery preparations and operations is separate project and is not the subject of the present paper.

Preparation of Pits

This will depend on the plant itself but normally a 2′×2′×2′ (60cm×60cm×60cm) pit is made. Depending on the plant the depth may even go to 3′ (90 cm). This pit may be filled with good soil transported from nearby, also with a lot of manure to increase water holding capacity and the other physical and nutritional characteristics of the soil in each pit so as to help the survival of the plant. One of the greatest problem is to make the seedlings survive to become trees and special attention has to be given to this point.

Planting of Trees

Planting should be done carefully, preferably in the late afternoon but whatever time the planting is done, adequate watering should be given at the same time. In some pits especially on some sloppy areas, mini-catchments should be developed for each tree as, during the rainy season, additional water can go to each pit as they are filled in with soil and some manure.

Maintenance

1. The maintenance of trees is a very difficult task. It suggests that the performance of 'Vanmahotsav' has been very dismal and that the normal survival may be less than 5 per cent. It is obvious that some technology has to be developed to increase the survival rate to almost 100 per cent. Trees should be planted in places where they can be taken care of and must be made to survive. *This operational research on the development of earthen pitchers is one aspect of the development of technology that will help in this direction.* Experiments on the following aspects have been done and are in progress for a suitable pitcher which will allow water to seep into the plant and which will need to be filled, say once a week, manually or by other means. Normally a pitcher of about 1½ litre capacity is to be put close to the plant in such a way that trees/seedlings are not damaged.
2. Experiments are in progress for the using of different clay, different baking period and with also different fuel to be used by the potters. For this purpose one in the village Khadambe near MPKV Campus who has baked such pots has been chosen. The data available gives a variable degree of seepage in this experiment. Seepage rate has been ascertained in slopy soils under field conditions.

3. The main object of developing such earthen pitchers and technology is to develop an ooze method of irrigation parallel to the technology of drip irrigation. The ooze technology utilizes local materials and men, and requires hardly any investment. If found suitable with some improvements, it may be recommended for large scale adoption.
4. The putting of a cement water mixture on the upper quarter of the pitcher's outer surface to reduce evaporation is one of such methods being tested. The lower three-quarters of the pitcher is in the soil. The study of the seepage rate which would permit the tree survival and which also reduce amount of labour and input for water, is being made.
5. The planting of trees and giving them proper environmental and soils and in short, taking care of each one of them requires some operations which cost effort and money. However, there is hardly any alternative in such areas, if we have to make the survival rate higher. The survival rate can be increased more speedily when unexpected rains come during the summer months or late monsoons and these will help reduce the cost and effort in filling the pitchers.
6. This note does not include protection from cattle and also from human beings. These are also crucial and need very great attention.

APPENDIX-VI

Dignity of Labour Camp for Land and Water Development

M. L. Dewan

Address to the Students Explaining Policy

My dear students, sons of the soil, builders of India and Maharashtra,

Today is a happy and proud day that our camp on "Dignity of Labour" has started. Your presence here today is proof that you decided that there is a dignity in labour, in working with ones own hands. We know that small efforts collectively means big efforts. The soil that you remove from one place to another, the ponds and dams you are building, the water you are storing, the trees you are planting, and compost pits which you are preparing, are the basis of agricultural development in this part and everywhere. We cannot continue to see barren hills and soil erosion – soil and water being lost before our very eyes. Either we have drought or floods, sometimes one followed by another. We should not blame God, we should blame ourselves. It is God's graces that we are getting monsoon rains which gives us an average of 500 mm – 750 mm which is still more than 2/3 of the land surface of the world gets. It is for us to make use of this resource and manage it properly.

When these dams are built, they are small, but each of them stores so many cubic metres of water. The trees are the life givers, their shade, their products and above all their slowing the runoff, prevent floods.

Let us dream together of clear streams of water without sediment. Let us dream that the streams will be perennial, having water for all or most of the time. It is happening in many other parts of the world with the same amount of rainfall. It is also possible here but we have to work for it now. There is no time to be lost.

Let us work so that various parts of our eroded hill areas be planted with useful species of trees, with quick growing fuel trees, with fodder trees and with fruit trees.

We have to stop burning cowdung. We have to make use of it. The fields need nutrient elements – nitrogen, phosphorus and potash which are in the farm manures, and many need organic matter plus something more. We can get the energy from the cowdung for getting *gobar* gas (biogas) and we can still use it for our fields and for our agriculture. We need mass efforts to have mass effect. Let us start now.

Planting of trees is important but still more important is to take care of these trees. Each one of them is like an individual who will flourish, with love and care. Let us care for each tree which we plant now. Give them good soil, if there is none around put some manure in it and give them adequate quantities of water when there are no rains. We would need storage of water for this purpose and also canal water. Let us plant hundreds, thousands and millions of seedlings soon. Now is the time, we cannot wait.

We have so much of waste material which we are not putting into use, which we are not recycling. Let us use every bit of farm waste including the straw of any other field crops as well as horticultural crops waste. Let us use every bit of our waste from colonies, villages and townships. Believe me, there is a lot of such waste. We would be benefited from both sides: one, we make use of waste material for improved agricultural production; and secondly, we reduce the pollution caused by such waste material. There is a scientific way of preparing compost. We would like to show you and let you prepare the pits in a scientific way, under the guidance of people who have developed these techniques and then work it out for your areas. You will work under the guidance of the Professor of Agricultural Chemistry and his associates. My request is that each one of you, as a son of the soil, should be familiar with the scientific preparation of compost. When you go back to the towns and villages, encourage such compost pits preparation with your farmers and others. All the above efforts are major inputs for any sound land and water development programme for any unit of land. Please do it on the MPKV Campus, on a large enough scale and this is the time for action. The objective is to grow not 20 per cent more crops but two times, three times, five times of the present level of production, in some cases more than that. This improvement is fantastic but it is practised in many parts of the world. I have seen them, witnessed them and participated in these efforts. Let us do it here, now and today.

Your region, your Maharashtra, your country will be proud of you, if you will work hard, sincerely and with devotion, so that you can be real builders of today's and tomorrow's India.

God bless you all !

APPENDIX-VII

Address to the Jubilee Session of the X Congress of International Society of Soil Science

*M.L. Dewan**

Dear President and Fellow Scientists,

On the occasion of this Jubilee Session of International Society of Soil Science, we bring you greetings from the soil scientists of India and from the Indian Society of Soil Science. India has over 1000 soil scientists, spread in the Government of India's research, teaching and extension organizations such as Indian Council of Agricultural Research and Indian Agricultural Research Institute, including All India Soil and Land Use Survey Organization of which my colleague Dr. R.S. Murthy is in-charge. Soil scientists are also working in 20 Agricultural Universities and State Departments of Agriculture and Soil Conservation. Those of you who were in February 1971 in India attending the International Soil Fertility Evaluation Symposium ably organized by Dr. J.S. Kanwar, then President and Dr. T.D. Biswas, Secretary of our Society, may recall the enthusiasm

* Director, Soil Institute of Maharashtra (India)
(Representing India and Asia on this Jubilee Session on 17 August, 1974, Address at the Convention Hall of the Hotel Rassia in Moscow to over 2500 Soil Scientists gathered from over 50 countries)

of our Indian soil scientists. Although we are sorry we are so few here amongst you, we bring you greetings from all of us.

After an excellent, and up-to-date review of the achievements of Soil Science by our distinguished President on the opening day of the Congress, I dare not go into a similar review, specially of what is more applicable to the developing countries. However, I intend to put before you the challenges which we, the Soil Scientists of India, are facing:

1. The first task and challenge is that we must increase the yield of crops per hectare. In many Universities and research organizations, we have got even 7 tonnes of wheat per hectare and 11 or in one case 14 tonnes of paddy per hectare by an optimum combination of soil, water and crop management practices. But it is not the plot which will feed our large population but the large and small holdings. The technology transfer to our small and medium farmers in addition to the few large farms we have, for real doubling, tripling and even quadrupling the yields consists of caring for every square metre of land, for every drop of water and for every plant.
2. Our second task and challenge is soil and water conservation, control and management. We are part of the "Monsoon Land of Asia" where we have the driest and wettest places in the world. Another characteristic is that either we have too much rain or too little, the so called average rainfall is achieved once in 3, 4 or even 5 years. So, we have either floods or droughts. We have built large dams and structures at a high cost of money and time, but we need more and more of small dams. We have to reduce losses of water in seepage and evaporation, do lots of land levelling, and provide water according to the needs of the crop. The recently started Water Technology centre at the Indian Agricultural Research Institute is spearheading the solution of this problem. We are doing water conservation

at arid zones and dry land farming research and development stations.

3. Another very serious problem and challenge is the provision of inputs: Demand has developed much faster than what we can provide by way of inputs such as seeds of improved varieties, fertilizers, pesticides and water about which I have talked. Our fertilizer factories are not able to cope with the demand not are we able to import enough fertilizers.
4. Our problem and challenge number 4 relates to the recycling of wastes and provision of organics. Recently, we have started a large scale campaign in "Scientific Compost Preparation". In an Agricultural University in India, where I have the good fortune to organize a "Dignity of Labour" Camp with 500 students who prepared amongst other works a large number of compost pits, we have lots of farm waste, farmyard manure, industrial waste, also human waste and town and city refuse etc. We have sewage plants in a few cities. The bones of our animals are also being underutilized and some even exported for profit. So, we have a large task ahead as regards organics and recycling of wastes.
5. Our challenge number 5 is solution of our soil problems. In a land area of approximately 330 million hectares, our soil distribution is :

   Indo-Gangetic Alluvium (Entisols and Alfisols), Red Soils (Alfisols and Inceptisols), Black soils (Vertisols and Inceptisols), Laterite and Lateritic soils (Alfisols and Oxisols), Desert soils (Entisols and Aridisols), Miscellaneous such as Mountain Soils. Salinity, alkalinity, waterlogging and erosion (both water and wind) and in some cases acidity constitute the major problems of our soils.

   Our soil scientists are playing an important role in solving these problems. However, we still have to work harder, to find better solutions and to speedily apply the measures of control and improvement which we already know.

6. Our problem and challenge number 6 is our large cattle population, over 500 million cattle, far too large and far less productive than efficient cattle. This high cattle population also highlights the conflict, we have between Man, Land and Cattle. This conflict can be solved by reducing the number of cattle, and by growing adequate fodder and by returning the manure to the soils *via* compost pits. We are doing this in a small percentage of our lands. We are burning most of our manure, as we are short of fuel for our masses. Kerosene is getting scarcer and costlier. We are not generating enough electricity. We have not yet evolved economic measures to harness solar energy for cooking. Our forests are depleted under the pressure of our people and cattle. We have, therefore, initiated a large scale tree planting programme of the fuel and fodder species and some green manure species, but this aspect of farm forestry including woodlots, shelter belts, and other plantings are essential for the improved used of our so called wasteland.
7. Although we have a number of problems, some enumerated above, we are also blessed with our land resources, water resources and solar energy of which we have a lot, almost all throughout the year. We have lots of people and we have embarked on a large family planning programme to control our population. We have to transform the energy and power of our people and to utilize this energy including Youth and Student Power of which we have a representation of over 120 million youths between the age of 15 and 25, for teaching them the dignity of labour and for development of our resources.

Dear colleagues and friends, the nature of our problems, is common to many countries, specially the developing ones, in the tropical and subtropical areas. However, the magnitude of problems we are facing is stupendous.

Our soil scientists are playing their part in solving these problems. We need to do more of all this. In these efforts, we have appreciated the collaboration of the soil scientists from other countries, from our distinguished President (Dr. Victor A. Kovda) who is also a member of our Indian Academy of Sciences, and has advised us often, also the collaboration of other colleagues from USSR, from FAO/UNESCO from USA and many other countries on a bilateral and multilateral basis. We welcome further such collaboration. In order to highlight our problems, challenges, and get the benefit of your views and experiences, we have brought an invitation from the Indian Society of Soil Science, to hold a Congress in India.

As the next Congress, we understand, is going to be held in Canada, we would invite you to hold the one after that hopefully in 1980/82. We have a variety of soils and we could show you these soils their problems, use and management and exchange experiences. I am sure you would all agree that the Soil Science Congress is not merely a gathering of distinguished scientists to exchange scientific views, but also to help mankind at large in improving agricultural production through the wise use and management of soil and land resources.

We, the soil scientists of India, therefore, consider it an honour and privilege if the International Society of Soil Science would accept our invitation. We are a poor country but we will welcome you warmly and heartily in our midst.

To end, I would like to add a personal note. Twenty years back almost to the day, when I first came to the USSR, on what was called a one man Indian delegation, I was struck by three things : the beautiful and imposing Moscow State University then newly built; the magnificent metros and above all the warm hearted and enthusiastic people and especially the soil scientists in all parts of the USSR which I visited. Now twenty years later, I am struck by the progress that Soviet Soil Science and the Soviet people have made in these years. We wish to thank all the Soviet Soil Scientists for their kindness and hospitality. Also our greetings on the Jubilee occasion from the Indian soil scientists to all the soil scientists gathered from all parts of the world.

> (The XII International Congress of Soils Science was held in New Delhi, from 12 to 20 February, 1982 – for the first time in a developing country.)

APPENDIX-VIII

Manpower Planning for Efficient Utilization of Agricultural Graduates in India

Address to VI Agricultural Universities Convention, Poona on March 15, 1975

*M.L. Dewan**

One of the key problems in rural development is provision of the trained manpower to create and staff the institutions necessary for servicing to the farmers' efforts to increase production. The major effort by the Government of India and the State Governments has been to develop and expand agricultural universities reflecting importance of these needs. As educational institutions expand, question continue to arise as the efficiency of training which they provide, the nature and productivity of the activities undertaken by their graduates and hence, as to the level of returns to educational investment. There is also concern as to the extent to which rural people themselves may have access to the institutions of rural development.

In order to efficiently utilize agricultural graduates for promoting agricultural development in the country, proper manpower planning is required. In this connection, it would be relevant to summarize results of a joint study conducted by the USAID/Cornell University reported in April, 1974. This study

* Mahatma Phule Krishi Vidyapeeth, Rahuri

reports implementation of all degree graduates including B.Sc., M.Sc., B.Vet.Sc., M.Vet.Sc., B.Tech. and M.Tech. and even Ph.D. by the G.B. Pant University of Agriculture and Technology for the years 1963 to 1971. The questionnaire based on 35 per cent of the degrees awarded, summarizes that only 10 per cent of the agricultural graduates have gone back to the farms and continued with the profession of farming. Rest of them have been in the jobs given in the following categories.

1. Non-university teaching in the intermediate schools, middle schools, high schools, as instructor, teacher, or demonstrator etc.
2. University research extension in various agricultural colleges either as Supervisor or Agricultural Officer (37.6 per cent).
3. University teaching in agricultural university as Instructor, Demonstrator, Assistant Professor, Associate Professor or Professor.
4. Government of India's Corporation such as National Seeds Corporation of India, Food Corporation of India, certain banks etc. as Officers, Inspectors, Assistant or Agricultural Officers etc.
5. Government of India's Research Institutions such as Indian Veterinary Research Institute, Indian Council of Agricultural Research, National Sample Survey etc. (Both 4 and 5 : 13.7 per cent).
6. Military service in the Indian Army as Officer-in-charge of the Military Farms or the Camp Officers (1-2 per cent).
7. State Governments in the various departments like Extension, Soil Conservation, District Agricultural Officers, Village Level Workers etc. (11.9 per cent).
8. Private business such as ESCORT, Indian Potash Ltd., Massey Ferguson and own business of the Agricultural graduates (8.1 per cent).

The percentage in each one of these if given also against this category. One more significant finding was that Bachelor of Science

(Agriculture), graduates, the highest paying positions are in private business or farming. In this category a graduate can expect to earn over his life time 1.7 times the earnings of graduates employed in Government of India Corporation, 1.7 times those employed in University research and extension and 1.5 time in State Government. Some policy recommendations can come out of the study as follows :

1. Expansion of programme of village level workers;
2. Establishment of Institute of Rural Development and Education;
3. Initial employment of the agricultural graduates by the Universities at an interval of 3-5 years.

I have a few suggestions to make in this subject of manpower planning.

To develop youth corps where agricultural graduates should work either in extension and development corps, education or health corps in the rural areas of India.

(a) Assisting through increased practical experience of the graduates and even undergraduates.
(b) Reduction in unemployment problem in the early years after completion of the studies.

Such a Youth Corps in India could be developed on the following four lines:

1. Agricultural Development Corps,
2. Education Corps,
3. Health and Family Planning Corps,
4. Rural and Urban Construction Corps.

These Corps will take also agricultural and other graduates as unemployment is a major problem. The pilot project undertaken in the Mahatma Phule Agricultural University for the Dignity of Labour Camps mobilizing students power for the construction purpose gave good results that students built up ponds, prepared compost pits, planted a large number of trees for soil and water conservation in areas. Students were both graduates, under-graduates, high school students both boys and girls.

The Youth Corps could first be developed as voluntary but may need to be obligatory. Huge organization is needed for the purpose, which could outline the conditions under which youths have to work for the rural India.

Agro-Industrial Operations

On the basis of experience of agro-industrial complexes in Bulgaria which have been considerably successful, agro-industrial operations are proposed for increased employment of agricultural graduates. The agro-Industrial complexes in Bulgaria included large state farms, also cooperative farms. The average size of such farm is over 24,000 hectares which is possible because of the socialization of agriculture. Although such large farms may not be fully applicable in India, the following specific task can be undertaken by the agricultural graduates and other educated rural youth.

1. Work on comparatively large farms both irrigated and unirrigated and their management.
2. Small scale industries such as horticultural industries like Tomato pulp preparation, Papine from Papayas, preparation of milk products from milk, poultry industry etc.
3. Seed production processing and its marketing of quality seeds to the farmers. As the demand of quality seeds is high not only by the progressive farmers but all farmers, their marketing is extremely essential. In practice the quality seed production and marketing is not in the hands of the people who are trained in seed production and marketing and does not always give suitable results.

4. Recycling of organic waste – In view of serious shortage in organic fertilizers recycling of organic manure is getting more and more important. There is considerable scope for agricultural graduates in this field for undertaking high speed mechanical or semi-mechanical composting by using rural and urban waste, some handmade paper plants, establishment of gobar gas plants etc.
5. Advisory services, agro-services in supply of inputs and marketing of produces can be given greater impetus by the agricultural graduates (This has already been in force in several States).

However, all the above tasks require considerable degree of devotion and hard work by the agricultural graduates. Many graduates, however, are looking for white collar jobs in teaching and research that has comparatively shorter hours of work and greater amount of leisure. This has to be changed. The graduates must be bold to take some risks, and show determination to identify with the rural population and live in the villages. In this process, they can help to improve living conditions of the villagers. This should be a strategy for agricultural graduates.

Modification in Curriculum

For overall manpower planning for efficient utilization of agricultural graduates, it is also proposed that out of four years, the last year should be compulsory for training on large farms, seed farms, research institutes, or with agricultural development activities, in the rural areas.* Unless this one year training is not satisfactorily undergone, the degree should not be granted. In other three years, it is proposed that at least four months should be devoted for fieldwork exclusively, 8 months for normal teaching, laboratory and extension work. An emphasis should be given to

---

*. Working with small farmers.

the dignity of labour. The work should be done manually by all the students, to develop confidence in themselves. A greater emphasis should be given to the schemes such as "Earn while you learn". Those who do not fit into this, may be sent out.

Each agricultural university must have a dynamic cell to give consultation to the needy emerging graduates. Their placement with large farms attached to the University and the University's adopted villages may be considered. These could serve as training ground for the newly graduating students.

Service to Small Farmers

The survey quoted in this note shows that only 10 per cent go back to the farm. The survey as to why agricultural graduates are not going back to their farm indicates as follows:

1. Farm size of certain joint families is small and either have no irrigation or very limited irrigation and cannot absorb another member.
2. In case of large irrigated farms, farmers have idea and expertise in the agricultural technology and the agricultural graduate who is trained does not increase efficiency of these farms. So the students of these families maximize their joint welfare through non-farming employment. The marginal benefit from having him solely on the family farm is less than if he is employed elsewhere.
3. Employment in one recognized institute, Government Organization, University or even reputed private businesses are a source of prestige influence which is appreciated by the family of the agricultural graduates.
4. Although not employed full time, on the farm the students or graduate can still act as a valuable consultant and information source for the family farm.

A meeting of Agricultural Graduates' Conference held in Nagpur in January 1974, has pointed out various factors for unemployment or underemployment and I summarize, the breakthrough in agricultural development, can be achieved by allocation of more money to the colleges and research institutes working with small farmers and rural development. A greater response to the needs of the poor farmers will require entire reorientation of our manpower planning. This will in turn need a larger number of agricultural graduates, each of whom can be assigned an area of operation. For example, out of 60,000 agricultural graduates, 20,000 could go to the task of agricultural research and teaching, 30,000 could be as agricultural advisers and extension workers, and remaining 10,000 could be working actually on their farms.

In the words of Robert McNamara, Ex-President of the World Bank has indicated "*the ablest administrators should no longer be reserved exclusively for the urban sectors.* Top engineering talents must be devoted to designing low cost solutions to the problems of small farm irrigation. Young graduates can be motivated to take on the *problems of rural poor and be adequately rewarded for solving them.* Educational institution must recognize that the training in practical skills in the college is important. In the developing countries ratio of Government extension agents to farm families should be about 1 to 400. In poor countries, such as India, it is on an average 1 to 8,000. And only a small fraction of these limited services is available to the small farmers. It is the deficiency of resolve to do more for the small farmers who desperately require them. There is scarcely a single developing country which does not produce too many lawyers and too many history graduates but there is no developing country which produces enough extension agents." And it is now the time to begin.

Part-II

Plans of Operation for Increasing National Efficiency in India (1965)

9 Problems of Development and their Diagnosis

India is in a critical stage of development, and consequently and necessarily, changes are accompanied by problems. The problems are of rising needs of an increasing population in the political, social and economic environments which have not yet achieved an equilibrium. There are ethnical, religious, language and even cultural differences, but these are all a part of society and in fact, play a significant role in development. However, India must develop rapidly and all the impediments only represent a greater challenge, and a hope of the development of a greater determination on the part of the Indian people to face the challenge.

It is also my strong belief that the natural and human resources of India, if properly developed and harnessed, will be sufficient to meet the needs of the present population of India. For this purpose, development must not be regarded as a mechanical process which has to be put in motion. It is a matter of organic growth – the process of allowing and encouraging people to meet their own aspirations. It covers the entire spectrum of a nation's life and moves round a central pivot which is the national will. It involves basic changes in attitudes and values, the adoption of new habits of thoughts and towards increased material welfare but also the least possible disturbance of the moral and spiritual values which the country cherishes. Development involves close inter-relationship between the various sectors of economic and social life. In India, where over 75 per cent of the population live on the land and derive their livelihood from agriculture, the key sector is obviously agriculture.

- Agriculture must provide adequate supplies of food for the whole population and wood and energy for fuel and housing. Agriculture in its widest sense includes the growing of many products of soil and water, i.e. it includes forestry and fishery. All agriculture is based on soil and water. Soils can be eroded by wind and rain and steps must be taken, including conserving existing forests or creating new plantations to prevent erosion.
- Secondly, soil needs properly distributed and regulated supplies of water to yield its best. Water, therefore, must also be conserved and controlled.
- It must supply raw material, i.e. cotton, jute, sugarcane and wood, etc., to provide the base for industrialization.
- It is the main source of investment capital, foreign exchange and manpower.
- Agriculture must provide a market for the budding industries and release labour for industries.

Crash programmes for industrialization are politically very attractive to those impatient for change, but these programmes will not be able to meet the basic requirements for economic growth, unless accompanied by speedy agricultural devlopment programmes.

There is no conflict between agricultural development and industrialization; they must go hand in hand. However, in India as in many other developing countries, *it is agriculture that has to produce the leverage for economic growth.*

What has been summarized above is the background of the needs of the people of India. These needs are not being met adequately because of the complexity of country's problems. A start has, however, been made, and the super structure of economic development is being built, but this structure needs sound and strong pillars. What are the pillars needed by India? How exactly can the foundations for these pillars which are essential for a strong structure be strengthened?

Some thoughts and summarized lines of action are given in the following pages. The six pillars for India's structure of economic development, in my view, are as follows:

I. Increase production – both agricultural and industrial.
II. Build individual character.
III. Improve national institutions.
IV. Education for all.
V. Control human population and carry out effective family planning.
VI. Improve and utilize fully cattle population.

All these six pillars are needed, although one may be more important than the other at a certain stage of development. Plans of Operation indicate an outline of ideas on how to construct and strengthen these pillars and how to construct and further vitalize the building of economic development for a fuller and more satisfying life for the large mass of people in India.

A paradox of idle manpower and natural resources side by side, with massive capital needs, poses a great challenge. As many economists have said "By far the greater part of a country's real capital structure consists of objects that require local labour and local materials for their production and construction". For example, many rural areas of India, like most of Asia and Africa, are in urgent need of better roads; irrigation canals, schools and health clinics, which would contribute directly to improved standards of living and require virtually no expenditure of foreign exchange. Yet the means to produce these improvements are now lying idle on most sides, landless laboures without jobs, inadequately used tools, indigenous raw materials and undeveloped skills. In the cities there are sewers to be dug, streets to be widened, fresh water systems to be introduced and above all, slums to be cleared and places of relative dignity, comfort and good feeling to be built.

Idle men and idle tools must be recognized as miserable, unforgivable, unnecessary insults to human dignity and a brake on national growth. This state of affairs can and must be changed. A large number of individual projects, with definite Plans of Operation must be undertaken within the framework of a programme and priorities. Uptapped human resources and skills will enable us to work towards a goal of resource development, human dignity and increased national efficiency.

(*P.S.*: Although these pages have been written about 15 years ago, most of it is equally valid today and are being reproduced as food for thought and fodder for action for development.)

In the following pages no attempt is being made to diagnose the ailments of a country as diverse as India. The diseases are several and some are in the advanced stages; however, there is a national will strong enough to cure many or all of these ailments. The time for further action has come and some of these actions are already being taken. But these actions are like that of the doctor who gives several different medicines to his patient, with a hope that one of the other may lead to a cure or lessening of the pain. To cure any disease, the diagnosis is important and an effort is being made to put some thoughts on the subject in the following pages:

(i) The basic question the country must answer is: are we going to run the country's affairs on a federal basis or a centralized basis? Or is it going to be a combination of the two? The way India's constitution is framed is logical, but is in actual practice greater decentralization needed. The lack of clarity in some cases greatly retards the speeding up of some actual action programmes by the State Governments.

(ii) Although there is every reason to be proud of the Government which is the greatest working democracy, the government machinery is still being run on the lines of the British colonial system, which is outmoded. In the so-called Secretariat System of Government where the responsibility is in the hands of administrators who, not being technicians cannot take full responsibility, nobody takes the blame. Responsibility is divided and dispersed in such a diffused way that ultimately no one in the Government bears the responsibility, except a few Ministers. One basic difficulty is that there are too many levels and too many people who keep the file in the Secretariat. The technicians usually have no or very little responsibility and authority. Often they act merely as advisers. *The men of action are told only to advise and not to act, and this is one of the greatest drawbacks.* This is slowly changing, however.

Technicians are wedded to their technical field for life. On the other hand, administrators move from Agriculture to Defence, from Finance to Internal Revenue and cannot build up a sound and efficient administration in a department.

Technicians, when they are given the responsibility, will not need clerks to give them notes and tell them what to do. There will then be a surplus of clerks who must be moved to some productive work in agriculture or industry.

(iii) Another basic problem is that of population and it is a vital one. The Government is taking a lot of determined action, but let there be no illusions, the effect is not going to be felt immediately. As the famous economist Gunnar Myrdal said in delivering the McDougall Memorial Lecture to the 13th FAO Conference in Rome, November 1965:

"Fortunately, the road seems now to be opening for a much more determined policy of spreading birth control among the masses.

"But we would deceive ourselves if we counted on a rapid decline of fertility. Also the relatively greater number of children, which is a reflection of the higher fertility would not always and immediately improve income per head and living levels, by lessening the dependency burden in the families, but it does imply that the birth curve in a nation would not be lowered as fast as if the age distribution were a more normal one, adjusted to a more stable population. Finally, the effects on the labour force are delayed for fifteen years and will be very inconsequential for a whole generation. The future workers are already born or will soon be born."

We talk of food for the growing population but necessities like clothing, homes, education and health needs for a growing population also have to be met.

(iv) What is the role of religion in modern India? We are a secular democracy and therefore, we profess that religion does not play any part in actual state of affairs. However, in actual practice, in several parts of the country, this statement is not true. In many towns a priest is, for example, and *ex-officio* secretary of the Town Council. Mixing of politics and religion leads sometimes to very unpleasant and unhealthy trends. These must end. The Government should also give more attention to the question of education which tends to remove a lot of prejudices. Decisive policy and action by the Government need to be taken.

These are some of our basic problems and these must and can be solved with determination. What is the solution? Work on several fronts in a planned programme, but with certain projects within the framework of a programme. The concentration of efforts should, however, be, in my view, on a few well-selected fronts. The Plans of Operation should be drawn up for these fronts and then the means established to execute these Plans.

India's Sixth Five-Year Plan (1980-85) has been presented and approved by the Parliament. It is a bold Plan. It would be useful to quote below the "Conclusion" under "Development Performance" and the challenges for the Plan period and beyond.

These comments and conclusions on the development of the Indian economy, given by an unbiased economist in the year 1981, may be of interest to the readers.

> "In the eighties, the economy is faced with a formidable task of maintaining and accelerating the tempo of economic growth in the face of a sharp deterioration in the international environment. The ever rising prices and highly uncertain supply of imported energy can disrupt the implementation of development plans. Nevertheless, the fact that the economy now enjoys the advantages of a high savings rate, a developed skill base and substantial degree of self-reliance provides a valuable cushion to absorb external shocks. If these advantages are to be turned to good account, we need to learn certain important lessons from our past experience."

With regard to growth and modernization, the first lesson is that we must mobilize resources to maintain and increase the tempo of investment and protect the size of the Plan both against inflation and external disturbances. The second is the paramount need to improve efficiency in key infrastructural and industrial sectors. The third is that we must reduce the variability of the growth rate and ensure that weather abnormalities and inflation do not lead to undue cutbacks in public or private investment. The fourth is the need to extend the agricultural revolution to all areas and farming systems and in particular to ensure that incomes of the poor are raised in the process of agricultural development. The fifth is the need to ensure a vigorous expansion in exports and a rapid increase in domestic production of oil, steel, fertilizers and vegetable oils so as to restrain growth of imports in order to maintain viability of our external payments. Sixth, we must develop effective domestic substitutes for imported energy so that our dependence on imported oil is contained within reasonable limits. Finally, there is an urgent need to revitalize the family planning programme so as to bring about a substantial decline in the birth rate through the voluntary adoption of the small family norm. Success on all these points is essential if the overall rate of growth is to be raised, as it must be, well above the levels reached in the past.

As far as self-reliance is concerned, the economy today is in a stronger position than at the start of the planning era. Dependence on external finance has been substantially reduced and in many areas of production a high degree of import substitution has been achieved. But there are many unfinished tasks in this area, *particularly in the field of energy and technology.* The world economy is in a much more disturbed state since the mid-seventies than at any time in the past three decades and there is an atmosphere of confrontation rather than cooperation in international economic relations. In this situation the *objective of self-reliance needs to be pursued with continuous vigour.* However, self-reliance can no longer take the form of indiscriminate import substitution. There is a continuous need to replace imports in critical areas where there are sudden and sharp changes in prices and availability. But, as the complexity of the economy grows, import requirements will also increase and in order to finance these export earnings will have to be stepped up substantially. In the eighties, export promotion is as much a part of the drive for self-reliance as efficient import substitution.

With regard to social justice, what we have achieved is far short of what we aimed at. After three decades of planned, development, large segments of the population have yet to share in the benefits of progress or participate in the process of development. From the Fourth Plan, several programmes for assisting backward areas and weaker sections have been in force. Yet, judging by the statistics on asset distribution, employment and consumption, the impact seems to be limited. What is needed is a more effective implementation of asset transfer measures, such as land reforms, more equitable distribution of credit and a coordinated effort that enables the poor to join the mainstream of economic activity and provides them with an opportunity for advancement. This will require firstly, an improvement in their productivity and earning power in their existing activity; secondly, supplementary employment in new activities to use up any spare labour time; and thirdly, *training, credit and support systems to assist them in both their existing and new activities.*

Growth, modernization, self-reliance and social justice are no independent objectives. They are linked in that success with respect to any one makes it easier to achieve the others. The sustained growth of the past thirty years and the very considerable diversification of our economic structure that has taken place during this period constitute positive national assets for launching a more direct attack on poverty and under-development in the Sixth Plan. However, the enormity of the task should not be under-estimated, particularly in view of a sharp deterioration in the external environment. The nation will have to mobilize all its latent energies for a more vigorous pursuit of cherished national objectives of *accelerated growth, greater social justice* and a *modern self-reliant economy.*

Project Approach to Agricultural and General Development in India

A new approach for speedier take off

In spite of the efforts made by the last Five Year Plan, agricultural development in India has not been adequate. Yields have not increased much, total production has gone up slowly and per capita production is about stationary if not less than before and only different. On the other hand, the population is rising greatly, and there is no noticeable increased standard of living.

The efforts towards agricultural development are not producing adequate results. A careful analysis of the problems facing Indian agriculture reveals a complicated picture. Further study, especially a comparative one, indicates that:

(a) *there is not a dearth of well-qualified personnel in most subject-matter fields;*
(b) *objectives to be achieved and targets to be attained are often clearly laid-out;*
(c) *budget or funds* (national and even international foreign exchange) *are not the major bottlenecks in a great number of projects;* and

(d) *there is no shortage of enthusiasm if it is properly channeled.*

What is *lacking*, in my opinion, *is a proper, clear-cut organization and management with definite responsibility for each of the members of the organization.* The huge manpower resources can only be an asset when they are properly organized and geared to the objectives. If this is not done, the same manpower resources can be a handicap rather than an asset.

Any change or reorganization must place a great emphasis on the quality of personnel, quality of work, etc. The quality can remain dormant unless a challenge, responsibility and powers are given at the same time.

My proposal is for a more universal adoption of "Projects Approach" to the organization of agricultural development in India. Each State should have some projects such as:

(i) improved fertilizer a manure use for rice in Uttar Pradesh;
(ii) efficient use of water for irrigation from the Tungbhadra Dam in Mysore State;
(iii) rural road building in Burdwan district in West Bengal;
(iv) conservation of red soil of Hazaribagh district, Bihar;
(v) reclamation of ravines in western Madhya Pradesh;
(vi) improved vineyards in new areas under Rajasthan canal;
(vii) reclamation of the Rann of Kutch in Gujarat;
(viii) improving rural literacy in Aurangabad district of Maharashtra State;
(ix) drainage and reclamation of the soils of Amritsar district (or one Tehsil) in Punjab, etc.

The Essentials of such a Project

1. Each project is in response to a request by the State or district indicating the needs and urgency of such a project. Each state should centre, say, ten requests towards the critical problem now, i.e. "Additional Food Production"

and each State should give a list of problems and priorities for increasing food production in their States:
Immediate (2 years)
Short-term (5 years)
Long-term (10-25 years)
Each State may suggest two or more projects under each heading, but in a review and final selection, it may be decided to give real emphasis to one project in each of the above time periods in each field.

2. An appraisal and recommendation on feasibility (both technical and economic) should be given by a group of advisers or consultants.
3. A body like the Planning Commission should approve such a request. This approval should include the total funds allocated including funds allotted by the State Government and the Government of India.
4. A Project Manager should be selected (how and by whom are important details to be worked out) whose tasks it would be to take the responsibility for *one* and only *one* project. He would prepare a Plan of Operation, a plan which should clearly state the objectives of the projects and how one achieves these objectives, a work plan, details of staff, organization, sequence of operation, reporting, financing, etc. Project Manager's tasks will include action to ensure that the operation of these projects is coordinated with current Central, State and local progrmmes and administrative personnel, rather than be completely isolated from them.
5. The Project Manager will be charged with the extension and management of the project. He will be given adequate power, consistent with the responsibilities he is to discharge. He will obtain assistance from the local sources and facilities as far as possible.
6. For the selection of project manager, some guidelines would be desirable such as:

(a) he should be technically sound and experienced but should have at least medium to good administrative ability;
(b) he should have qualities of leadership and should demonstrate flexibility of approach; and
(c) he should be an *operational* type not an intellectual type and should enjoy taking responsibilities.

7. For keeping good project managers on the job and fully interested, there should be a good incentive and definite possibility for meritorious promotion so that one can see the future clearly.
8. Project managers will not be interfered with day-to-day work. They will submit a progress report, say every quarter to the Supervisor of the Centre (or the State Headquarters) when it should be possible to measure the progress of work against the Plan of Operation.
9. Technical and budgetary responsibilities rest with the project manager. The staff at his disposal will consist of well-selected technicians as given in the Plan of Operation. His personal staff will consist essentially of one secretary and a minimum of administrative staff.

(P. S.: The project approach has since been initiated and/or is going on in several parts of India in local, state, national and international projects and it is time to evaluate and further improve the application of this approach for development.)

Explanatory Notes

What is a Plan of Operation? What is meant by Executing Agency? What is the concept of Volunteer Action Group? What is Government Cooperating Agency? What is Special Project Allocation?

What is a Plan of Operation

These terms are going to be used and hence some explanatory notes should prove of assistance

A Plan of Operation is an agreement between the parties concerned in executing a project. It contains a detailed description of the arrangements for the execution of the project. As a "blueprint" for a project, the Plan of Operation describes who is to do what, when and how. As a guide, it lays down the conditions and methods of operation. As a control document it provides the criteria against which the parties concerned may measure the progress of the project. As a contract, it is a detailed statement of the obligations of the three parties involved in the project. Below is given an outline of a typical Plan of Operation.

I. Summary

Special Project Allocation:
Consisting of non-Government Bodies Contribution
Consisting of Government Contribution
Duration of Project
Executing Agency
Cooperating Government Agency

II. Purpose and Description of the Project

A. Purpose
B. Description

III. Work Plan

A. Participation and Contribution of...
B. Participation and Contribution of...
C. Participation and Contribution of...
D. Organization
E. Sequence of Operation

IV. Budget
V. Reports
VI. Revision

VII. Steps to be taken on completion of the Project

VIII. Signature

IX. Appendices

Appendix I : *Plan of Expenditure*
Allocation by... (Government)

Appendix II : *Plan of Expenditure*
Allocation by... (non-Government Organization)

Appendix III : *Total Contribution by...*

Appendix IV : *Schedule of Operation*

A. Manning Table for Professional and other staff training and Fellowships

B. Other Facilities:

(a) Building, Land

(b) Equipment

(c) All other services and facilities

C. Work Programme

Appendix V : *Relevant Maps or other documents.*

What is the Concept of Volunteer Action Group?

Details of this group have to be worked out. The unit which is, for example, being led by Shri Rajmohan Gandhi (Moral Rearmament Group in India) could be a Volunteer Action Group for the purpose of "Improving Individual Character". AFPRO, CARTE and hundreds other voluntary agencies and people are ready to help serve in food production.

The same action group could undertake other tasks outlined in these Plans of Operation, but this needs to be discussed further.

Other details are financing, sequence of operations and timetable (in a tabular form), organization (i.e. the definition of the tasks of the Project Manager and others) have still to be discussed.

The outline of the following Plans of Operation gives only the purpose and description. A great deal of time and team work will be necessary before other details can be finalized.

10 Plans of Operation

A Special Project aimed at further increasing Agricultural Production in India

Special Project Allocation

Duration:	Concentrated Effort: Ten years
Executing Agency:	Special Projects Agency (Volunteer Action Group)
Cooperating Government Agency:	Ministry of Food and Agriculture

For the purpose of increasing agricultural production, with special emphasis on food production, which is the greatest need of the decade in India, the Special Projects Agency undertakes, on an emergency basis, the Plan of Operation which is pledged by the Volunteer Action Group and for which the Government of India and the State Governments have instituted and undertaken considerable steps and measures and still recognize that Governments alone may not be enough and that farmers, people and non-Government organization can generate considerable resources in the agricultural development programmes.

Purpose and Description of the Project

The purpose of the project is to effectively assist in increasing the agricultural production in India, on a continuing and logical basis and with special emphasis on food crops.

In particular, the project will :

(a) Study (investigate) and identify the major bottlenecks

in agricultural production, and concentrate on the removal of these bottlenecks with a definite priority (Sixth Five-Year Plan has done this adequately).

(b) Tap the human resources which are unutilized or under-employed for agricultural development purposes, by initiating action groups of people of all ages.

(c) Assist in utilizing all arable lands in as intensive a way as possible and for such an intensification, the inputs must be made available and in time (irrigation, good quality seeds, fertilizers and manures, etc.).

(d) The responsibility should be distributed and large incentives be given for increased production to a large number of people, not only to farmers who are, of course, the most important, but also to cooperative state farms, school gardens, kitchen gardens maintained by the town and suburban population.

(e) Losses in harvesting, transportation, marketing and storage should be reduced to the minimum.

(f) Greater number of people for actual production, fewer people between producers and consumers. Cooperatives as far as possible.

(g) Project approach (see Chapter on Projects Approach to Agricultural and General Development in India).

Description of the Project

For the purpose of increasing agricultural production, the following action should be carried out or at least initiated. The list is not exhaustive but illustrative.

(A) Improve the *efficiency* of what is going on :

Through the Government

Increase the supply of fertilizers and manures

Better and more seeds through certified seed farms

Improved soil and water management

Better irrigation practices

Provision for supplies of additional tree seedlings

Diversification of crops

Establishment of additional fuelwood plantations so that dung is not burnt

Improved marketing

Reduced losses

Improved storage.

Farmers

Greater incentives–prices

Modify eating habits

Grow more vegetables

Grow more fruit

Grow more trees, especially for use as fuel to save dung, for building poles and for cash sales.

Manage and use soil and water efficiently and intensify their use as far as possible.

These and many others, although many of these are slow processes, are needed.

(B) New avenues, new approaches or partly new ways must be sought as follows:

1. Voluntary organizations

– Some action groups, local leadership, self-sustaining, self-propelling.

Each district should have a voluntary organization called "Volunteers for Increasing Agricultural Production".

These volunteers should have camps at one or more places in the district (1000 ha production farms). They should be run on military lines. Each volunteer gives his time as required and he gets food and certain pocket money depending on what he has been doing. In addition, at the harvest time he gets a share of the crop. Many of these volunteers may be for the peak season for sowing, harvesting and threshing.

There is a total of 393 districts in which 1000 ha farms should be set up. These should be called *Pilot Farms*, meant for demonstration and training, as these should be under the leadership of active and practical workers and a number of technical personnel or specialists.

2. All boys and girls of certain age groups should give a minimum of one year and preferably two years of their youth to the country:
   Boys, any one year from 16-20
   Girls " " " " 15-18

(These are pre-marriage years. However, a married couple can give one year from 18-25 years of age). A special chapter is to be devoted to this matter and will need a small group of practical and experienced consultants to prepare it in 2-3 months or more. Out of an average of 15 months in volunteer service,

normally, 9 months in agriculture, horticulture, forestry
3 months in military, sport and physical exercise
3 months in social service: roads, hospitals, schools, etc.

If considered more appropriate, two years would be requested, of which one year would be for agricultural crop production. Or classify into two: i.e. some for one year and others for two years. It should start voluntarily.

3. *Diversify local foods*
   - Grow more vegetables
   - Have a milk cow and use more milk and milk products
   - Grow more beans, pulses and leguminous crops
   - Fish where possible
   - Meat and meat products. Where acceptable this area

should be enlarged as far as possible to include poultry and eggs.

- Date palms where climatically possible
- Improve cooking methods and facilities.

4. Have private companies establish farms.
   Tata Steel, for example, should have, say, a 10,000 ha farm for producing all the needs of the personnel at Jamshedpur, another one near Bombay, a third one at the Tata Chemicals Ltd., Mithapur, etc.
   All workers should have free holidays on these farms; 25 per cent of holidays should be used for agricultural production.
5. Spread good farmers into poor farming communities. This will have a great demonstration and training effect which will facilitate his absorption into the community which is always difficult. He should set standards which others should attain and should not himself lower the quality.
   In this connection, diploma holders of agricultural schools who have shown promise of being practical farmers and who will apply improved methods should be given land free or at small instalments and not be given white collar jobs.
6. Organize agricultural exhibitions at district, state and country level *regularly* once a year and have *field days* for a large number of farmers. Further utilize "religious fairs" fervour to demonstrate advances in agricultural production techniques and for practical purposes include demonstrations on tree growing and uses of wood, new and efficient wood burning stoves, improved methods of cheap house building.
7. Establish or strengthen commodity research stations and pilot farms for commodities such as:

Grapes	Sugarcane
Apples	Maize
Peaches	Beans

Pistachio	Pulses
Almonds	Olives
Sugar beets	Mushrooms
	Walnuts

There are already some for rice and wheat and a few other commodities. On pilot farms, include the planting of trees and fuel wood plantations.

8. The biggest industry in India can be the production of a second and sometimes a third crop by proper cultural, conservation and irrigational practices. It will,
   (i) increase agricultural production
   (ii) enable the fuller utilization of underemployed human resources.

India as a whole has a cropping intensity of 120 while China has about 200 (2 crops on every hectare of land on an average).

Plans of action should be drawn up for irrigated areas, partially or supplemented irrigated areas and non-irrigated areas.

9. We normally talk of increasing food production in the farms. A large number of people live in:
   (i) urban communities,
   (ii) suburban communities, in addition to a great majority being in, and
   (iii) rural communities.

Plans of action should be worked out as to how to increase food production in all the three sectors. The kitchen garden can play an important role, and even pots can be used to grow some food. It depends on how many times the pots or kitchen gardens are multiplied.

Also in rural communities, draw plans of action for:

(i) individual holdings
(ii) cooperative farms
(iii) collective farms
(iv) state farms

Let there be a competition between various types of farming and some special distinction for the outstanding ones, goals being in terms of production per unit area per year.

Plan of Operation

A special project aimed at building individual character in India

Special Project Allocation

Duration:	Concentrated Effort : one Generation (25 years)
Executing Agency:	Special Projects Agency (Volunteer Action Group)
Cooperating Government Agency:	Ministry of Education/ Ministry of Social Welfare

For the purpose of carrying out effective steps to build national character which is a summation of individual character (which has been deteriorating in the past years) and which is the key to all permanent improvements on the mental, material and spiritual planes, the Special Projects Agency should undertake, on an emergency basis, the Plan of Operation, which is pledged by the Volunteer Action Group, and for which the Government of India are considering effective steps through the Ministry of Education, Ministry of Social Welfare at New Delhi and similar measures through the States' Ministries of Education.

Definitions to be given of the following:
National character
Happiness
Material prosperity
Mental prosperity
Spiritual prosperity
(not to confuse one with the other)
Individual character.

Purpose and Description of the Project

Purpose

1. The purpose of the project is to stimulate the development of the character of all citizens of India who are responsible and not corruptible, to become the pillars of a great country, and who will each carry out with efficiency and reliability the job or task that each one undertakes.
2. In particular, recognizing that character is built in the early years of a child, like the foundation of a building; also recognizing that even though the foundation may be strong, the building can still be weak, the project will:

    (a) Study, find out and establish the reasons for the deterioration of national character, including methods for regenerating the will and determination to improve it;
    (b) Stimulate effective and practical measures to teach at an early age in all the schools the principles and practices of high character which is not corruptible;
    (c) To bring in the principles and practices of absence of corruption in everyday living in homes, in business, in offices, in the field, in politics, so that the growing children and future citizens do not get disillusioned by seeing that all they were taught in schools is not applicable to everyday life; and
    (d) Leaders of men, in political, educational and other fields should be persons who are not corrupt nor corruptible as they are the ones who set examples to others.

Description of the Project

Specific actions such as follow will be stimulated:

1. *At home*: guidance of mothers and parents to bring up the children properly and adequately and assist in building their character from the beginning. At Central, State, District and Thana, or Block level, organizations of enlightened mothers, few but devoted and volunteers: their task is to show how important it is to have a strong character, to be honest, to do one's share of work, to be truthful, to do one's civic duties, even to do voluntary work for national welfare and development.
2. *In schools*: good, selected, well-trained. moderately well paid teachers. The curriculum to be worked out carefully; and properly prepared text-books with practical exercises should be used. In studies, in games, and otherwise, the emphasis should be placed on strength to stand up to injustice, nepotism, unfairness, etc. Movies, trips, camps, establishment and running of school garden plots and school forests, which are useful for practical work to gain an understanding of the importance of various relevant factors in character building, should all be included where possible in school activities.
3. *In colleges and universities*: every student for one out of 2-3 summer vacations, must join training and practical camps (also give one or more years of his life to the country). If necessary, a law should make it compulsory that each youth should give one or two years to military and civil trainig. (Can we organize it? Yes, give 1-2 years to organize curriculum should be worked out).
4. *In business :*

   (a) that holiday camps should be organized at the important holiday resorts, for medium to big business men for a few weeks each, where they are invited and the importance of character discussed through movies, seminars, etc.

   (b) any suggestions to be incorporated.

5. *For offices*: Government offices—there is to be more responsibility given to each pers,on and there should be fewer people in every office, varying according to the need. There should be fewer steps for each task. (The project approach to actual development should be enunciated. To define the Project Approach see "A Special Project aimed at Improving National Institutions in India").
6. *For political and public office holders*: complete honesty and cleanliness in their jobs and dealings.
7. *For industrial workers*: holiday camps should be organized and planned. Courses be planned. or movies be shown indicating the importance of hard work and sound character, which should work positively and for increased production.
8. *For farming population*: offering them the leadership and incentive to work harder and increase production. The leadership should help to remove the bottlenecks or impediments for their increased production.
9. What are the theory, practice and economics of corruption?
   (a) Generally speaking, corruption can be defined as the inappropriate use of power or influence which somebody has because of his position. This misuse can be and is usually for:

      (i) immediate private gains,
      (ii) political party, and
      (iii) pressure of one kind or another.

   (b) Corruption has some other characteristics:

      (i) These involve usually the need to keep it (corruption) secret,
      (ii) There are usually laws in most countries against corruption. The lawmaker has decided that corruption is punishable, but often the borderline between proper and inappropriate action is not well defined, and

(iii) This borderline is often decided by custom. Custom varies from time to time and at different places.

(c) Why is the ground fertile in some places for corruption or the practices of corruption?

(i) When the public sector expands rapidly and public authorities take charge of commercial activities, or of new regulative functions (investments, foreign exchange control, fixing of rents, prices, allotment of houses, land, etc.).

(ii) When the social machinery and existing value concepts of a society are destroyed, or when a whole nation has to suffer hardship and starvation while individual profiteers can live in abundance and luxury, it is difficult to expect that the civil servant's sense of responsibility and loyalty shall remain unaffected, especially when he is underpaid and lacks security of tenure. "It is hard for an empty sack to stand upright".

It is difficult to require honesty when the economic prerequisites for honesty do not exist.

(d) How to fight corruption?

(i) Build individual character.

(ii) Simultaneously work on the economic prerequisites.

(P.S. : Recently (2006/07), I have prepered a book jointly with Artist Pradeep Nayak, entitled "Human Values — A Voyage from I to WE" one page Summary of the content of the book to present/reduce Corruption in our country).

Plan of Operation

A Special Project aimed at improving National Institutions in India

Special Project Allocation	
Duration:	Twenty years
Executing Agency:	Special Project Agency
Cooperating Government Agency :	National Efficiency Council

For the purpose of *improving national institutions, and improving the efficiency of a large number of people in India* manning these institutions and thus to serve a still greater number of people in India, the Special Project Agency undertakes, on an emergency basis, the Plan of Operation (which is pledged by the Volunteer Action Group) and for which the Government of India are considering of strengthening the "*National Efficiency Council*".

It is recognized that in India, as in many other developing countries, economic under-development is in itself largely a consequence of Institutional Under-development, (i.e. Institutions, either non-existent, inadequate; or improperly oriented to meet the needs of economic progress). Also, it is recognized that as *India wishes to enter the stream of economic progress*, she must be willing fundamentally to alter her institutional structures and that the institutional transformations called for will be deep, profound and far reaching. There are many institutional impediments to rural progress in India, such as :

(i) Institutions which inhibit the play of incentives (to work hard, to save, to invest, to innovate, to take risks, to acquire skills). These are:
   (a) Land Tenure Institutions,
   (b) The Credit System,
   (c) Pricing and Marketing System, and
   (d) Type of Social Organization, such as the joint family or the tribe.

(ii) Institutional factors which inhibit the development of

capabilities of rural people (development apparently depends primarily on the development of human resources and capabilities to give the opportunity for individuals to acquire meaningful experience while they are young enough to try new ideas and to learn from this experience and thus to develop self-reliance and enterprising and managerial skills).

(iii) Institutional factors which exhibit the development and utilization of science and technology:

(a) Extension service requires Government officials to have a "servants relationship" rather than a "masters relationship" with the farmers, to give them the technical assistance which they need rather than some administration.

(b) Institutions of formal education and research including agricultural colleges, extension or community development services need to adapt themselves to development roles. (Much more effective means of farmers' education and farmers' participation and indeed control must evolve if research and extension are to become truly at one with the agricultural development process. This is possible by demonstration in the field of tested methods, by bringing better farmers from other communities to settle among the poor farmers, show movies and use other audio-visual methods, also exchange of farmers, or travel if possible in large numbers to other areas).

Purpose and Description of the Project

The purpose of the project is to effectively assist in improving national institutions in a progressive way and especially those which are mainly responsible for accelerating the process of economic and social development. In particular, the project will:

1. Classify the type of institutions and study the reasons for the shortcomings and pitfalls in these institutions to meet the tasks and targets that have been set for them.
2. Plan, together with the appropriate Government, non-Government and even foreign assistance agencies, where required, the measures for improving such institutions in each of the important sectors of development.
3. Concentrate on the agricultural and educational sectors in the first instance and give broad outlines for industrial, health and other sectors.
4. For agricultural institutions sector, promote the :
   (a) development of agricultural education and training institutes;
   (b) development of research facilities;
   (c) development of effective institutes for agricultural supply lines and of storage, transportation and marketing facilities;
   In the first two aspects considerable progress has been made in the last 15 years.
   (d) development and operation of effective extension service, including development of incentives to, encourage increased production.
5. Recognize that any institution, for proper and efficient development, takes at least 10 or 15 years, and not 1, 2 or even 5, although significant improvements can be made in a lesser period, and so promote awareness of the need for time for any sound institutional development.
6. Give special emphasis to individual efficiency, the summation of which is national efficiency. The personal efficiency of an individual is development and the putting into practice of the best in a person, in study, work and play. It is based on four fundamental principles:
   (a) Good health
   (b) Right and positive thinking
   (c) Creative and productive work.
   (d) Good environment

Apart from these principles, the other key points in the art of personal efficiency are:

(i) All interest which is creative-interest is the mainspring of action. Action means effort. Effort of any kind is enough to create the beginning of the kind of personality that aims for efficiency.
(ii) Future success would be built on past and present achievements. We add to our existing self-confidence by developing that self-confidence within its own field and thus venture into fields not so familiar and taste the joys of experiments and risks.
(iii) Sense of values and sense of purpose.
(iv) Good memory.
(v) Good appearance.

Plan of Operation

A Special Project aimed at Education for all in India

Special Project Allocation	
Duration:	One Generation (25 years)
Executing Agency:	Special Projects Agency
Cooperating Government Agency:	Ministry of Education

With the purpose of promoting eduction for all, it is recognized that education is directed towards a fuller development of human resources of which there is an abundance in India. It is also accepted that these are under-utilized and that education will assist considerably in this process of full utilization of the vast human resources. The Special Projects Agency will undertake, on an emergency basis (as part of the War on Ignorance) the Plan of Operation, which is pledged by the Volunteer Action Group and for which the Government of India is considering reinforcing the Ministry of Education at the Centre and the Ministries of Education of each State.

(i) Education is ability to read and understand the recording of the thoughts and experience in books, accumulation of collective wisdom and the learning of centuries.

(ii) There may be some differences of opinion on the kind of education, on education methods, on the amount of education, on who should have the responsibility for education, on who should pay for it and on many other questions such as curriculum and syllabus, but there should be full appreciation of the need and value of universal education in India.

(iii) India must learn the following from certain developed countries:

  (a) Returns from an investment in education and research in all sectors and specially the agricultural sector, although very large, do not come quickly but often require a generation or more for full fruition.

  (b) The greatest gains in agricultural production usually do not come from an improvement in a single practice, but from the interaction of a suitable combination of improved practices.

  (c) Agriculture cannot advance alone. There must be simultaneous development of the industries which supply essential goods and services to agriculture.

  (d) Large returns for agricultural development are not realized until the farmers responsible for a high proportion of the national production have adopted the improved practices.

(iv) Education at all levels is important. More emphasis in India needs to be given to education, but this has been lacking so far due to :

  (a) shortage of funds;

  (b) shortage of qualified teachers. This shortage will limit the rate of progress, even when adequate funds are available.

(v) Education is of the following types:
  (a) Elementary or Primary School;
  (b) Intermediate or Middle School;
  (c) Secondary or High School;
  (d) Undergraduate College Education;
  (e) Post-Graduate University Education.

  In addition there is :
  (f) Education such as vocational agriculture and other vocational schools for trades and industries;
  (g) Adult Education which, although not included in formal education, should play an important role.

(vi) It is important that India encourages a fairly high percentage of her educated, industrious and pioneering young men to remain in farming and to ensure progress in agricultural development. Both the quality and quantity of training must improve in the field of agriculture.
  This will need careful revision of the curriculum of agricultural and educational institutes.

Purpose and Description of the Project

Purpose

The purpose of the project is to effectively assist and promote ways and means to have "*Education for All*" in India. In particular the project will :

1. Assist in assessing the short-term and long-term educational needs and plans at all levels of education.
2. Assist in establishing a "*Youth Education Corps*" (YEC) wherein a number of young men and women, between the ages of 16-25 will volunteer at least one year for teaching the uneducated and school children. As a result, a certain percentage of the YEC may take up teaching as a profession and take additional courses for teachers' training and other specialized courses.
3. Assist in establishing a "*Senior Education Corps*" of people

who have retired from Government and other services and are in good enough health to give their time, part or whole, for teaching purposes, especially in advanced technical institutions.

4. Strengthen the existing agricultural schools, colleges and universities in various parts of the country for farmers children and for assisting those who would like farming as a profession, or who service the farmers. In general, more practical schools are needed rather than colleges and universities.
5. Stimulate that in "Education for All" the important element of quality should not be ignored. The utilization of the intellectual potential of youth that have not been fully utilized and developed in elementary and secondary schools and promotion of newer methods of teaching should be given adequate emphasis.
6. Propagate *moral and spiritual principles* which India has held dear and be given great importance together with the scientific education. The heritage of India should be given its due place in the education method and syllabus; although it is recognized that these principles need reorientation to modern technology.
7. All education be aimed at :
   (a) Building character;
   (b) Practical works and underscoring the dignity of manual labour principle; and
   (c) Self-reliance by people of all ages.

Plan of Operation

A Special Project aimed at Effective Family Planning in India

Special Project Allocation:
Duration: One generation (25 years)
Concentrated Effort: 10 years

Executing Government Agency: Ministry of Health and Family Planning

This project recognizes that people can be a great asset when properly managed and also that there is a race between people and food production. (The population in India is going to grow barring catastrophes, even if birth rates fall with unprecedented speed. For example, even if the birth rate was suddenly cut in half, the population would still grow more rapidly than it did in the 19th century). There is no reason to expect a stationary population for decades. The only realistic hope is that the rate of growth can be brought to *manageable levels.* The population of India will probably be double it is now (1965). The real question is whether it will be in 25 years or in 60 years. The present Plan of Operation is aimed at making the population grow in a manageable, not in a chaotic way. The facts at present are that food production per head in India has tended to decline over the past few years. India has the good fortune of having a population which properly developed can be the greatest asset for its resources development. However, the important task at present is that production of goods and services should at least keep pace with population growth. However, there is no question that the multiplication of needs is also taking place and that needs are not stationary.

In India, there is an excess of population as compared to the food we produce, and for the purpose of carrying out effective family planning and thus keeping human population to a manageable level, the Special Operation is pledged by the Volunteer Action Group and for which the Government India are considering steps with a view to strengthen the Ministry of Health and Planning at the Centre in New Delhi, with considerable powers and finances, and other similar measures in the States.

Purpose and Description of the Project

Purpose

1. The purpose of the project is *to slow down and ultimately*

arrest the present high rate of growth of human population, and thus to stabilize the present population of India at a manageable level and in an organized scientific way and with emergency measures, as the *achievements of people and nation depend more on the quality and much less on the quantity of the people.*

2. In particular, the project will :
   (a) Accumulate the existing information and where necessary study, find out and establish the reasons as to why the population growth has been increasing at such a high rate, especially in the last 20 years or so. (Research in social and psychological aspects of family planning, job of the Lucknow Institute of Planning, Research and Action on Family Planning).
   (b) Stimulate and where possible *carry out the measures necessary to offset the need for increased population* in individual families and communities (e.g. more hands, more work forgetting more mouths, more food). This can be done by *giving old-age pensions; homes for the old;* in rural *sectors improved farm implements*, hand or animal drawn and even mechanization where feasible in areas such as Rajasthan, to offset the need for a large number of hands as desirable for certain farm operations, further strengthening of the inherent capabilities of women, age for marriage, etc.
   (c) Institute *steps for large-scale production of means for an effective control.* These measures fall into several broad categories:
      (i) Industrial-manufacture of the prophylactic, loops and other means for population control on a large scale, adequate quality and at a cheap rate;
      (ii) Sterilization and its equipment;
      (iii) Training of personnel to carry out required measures for an effective family planning programme.

(d) Effectuate means to promote the large-scale demand and use of the means and measures of family planning, e.g. :

   (i) distribution at a very nominal rate of the means together with instructions for their use;

   (ii) carry out voluntary sterilization programmes with increased stimulus and incentives;

   (iii) spread the programme through the media of mass communication including audio-visual measures, newspapers, special pamphlets, radio, television, etc.;

   (iv) large-scale general education of the masses will be an important step, but this will be the task of the Ministry of Education concerned.

(c) Coordinate and capitalize on all volunteers, official and non-official, regional, national and international directed to this important national and international issue of family planning and control of human population. In particular, utilize the unemployed and under-employed corps of educated personnel (including lady doctors and health and social workers) and the increased utilization of community development block offices and the village level workers and other similar agencies.

Plan of Operation

A Special Project aimed at Improving the Efficiency of the cattle population

Special Project Allocation :

Duration:	One generation (25 years)
Concentrated Effort :	10 years.
Executing Agency:	Special Projects Agency (Volunteer Action Group)

For the purpose of making the present cattle population of 200 million (1965) effective and efficient, which may involve considerable reduction in the first stages of useless cattle, the Special Project Agency undertakes, on an emergency basis, the Plan of Operation which is pledged by the Volunteer Action Group, and for which the Government of India is willing to institute appropriate steps (including starting and reinforcement of appropriate legal measures if required).

Purpose and Description of the Project

Purpose

The purpose of the project is to improve the efficiency of the cattle population in India. This may involve reduction of the useless part of the cattle population in the first instance and to increase the number and percentage of efficient cattle. Efficiency means the fulfilment of the purpose for which cattle exist. In particular, recognizing that cattle must be converted into an asset, that there are a number of impediments, for example, superstition and ignorance, involved in the question of cattle; also recognizing that they are considered sacred, this project will be specifically expected to undertake the following steps:

(a) survey the latest information on cattle wealth in India regarding number, types of breeds, their output;
(b) prepare a plan of operation for rational and efficient harnessing and utilization of this cattle population including more and better hay and pastures, more better milk and milk products, animal breeding and selection centres and veterinarian services;
(c) initiate a large number of factories in a progressive manner for utilizing carcass, hides and skins, bones fertilizers, glue industries, etc., in areas of potential raw material;

(d) prepare a careful programme of educating public opinion on:

   (i) triangular conflict between Soil-Man-Cattle:;
   (ii) meaning and history of the "Sacred Cow"; criteria for sacredness (cow for milk, bullocks for agriculture and transport) and the conditions which do not apply at present; and
   (iii) science and technology have modified conditions.

Education should be started from childhood where possible.

(e) all "*Goshalas*" should be cattle improvement centres;

(f) carry out scientific and etymological studies as to the effect of meat consumption in hot climates, especially with the latest techniques of refrigeration. (My theory: sheep and goats being small could be consumed by a smaller community). Cattle (cow or bullock) were consumable by larger groups, and if not consumed quickly could not be stored because of the hot climate and hence may have resulted in food poisoning in a number of instances. Beef eating was proclaimed as irreligious, because religion always produced a quick response from people, consequently this ban became a norm of life).

Estimated annual loss because of inefficient cattle is Rs. 1,700 crores which is practically the same amount invested in development (plan layout).

Pinjrapole = old cattle house (3,000 acres of valuable land) useful but needs careful unbiased evaluation.

(P.S. : Innovative work, investigations and research and action spearheaded by Prof. N.S. Ramaswamy, Director, Indian Institute of Management, Bangalore, during the last 5-10 years have made the beginning of an input and need forceful pursuit including an animal energy corporation and other developmental' and institutional measures.)

Plan of Operation

A Special Project aimed at Increasing industrial Production in India

Production can be either:

1. Agricultural production, or
2. Industrial production the task of industrial production is to produce goods which satisfy the growing needs of the people for a fuller living. Industrial production for houses and accessories, for roads, railways, airways and other transportation systems, for clothing, for social, cultural, recreational and other needs of the population.

Although this Plan of Operation should fully deal with industrial production. I am restricting myself to reproducing below the material written on "industrial development based on renewable natural resources" by a FAO technical officer in this field and which is of great relevance to the question of industrial production in India.

"It is fully recognized that industrial development must be given very high priority. However, there may be some differences of opinion regarding the types and size of industries that need to be established first. Sometimes, preference is given to the development of heavy or strategic industries, described as the manufacturing sector.

"Some years ago this latter viewpoint was shared by quite a number of policymakers, who took action and commenced a programme of industrialization based on large industrial units. There were some promising developments, but there were also quite a number of failures, especially in countries where not enough, or no attention at all had been given to the development of food and food product industries. This mistake retarded other industrial development programmes and often caused financial problems almost impossible to solve.

"Very large countries, when having access to exploitable deposits of ores, minerals, and oil, will naturally benefit greatly when starting to develop large industries based on these non-renewable natural resources. As these types of large industries are more capital and less labour intensive the amount of workers employed per unit will not be very high. Thus, only a small percentage of the entire population will benefit by receiving higher wages. However, the economic advantage lies in the fact that the country will greatly benefit in earning much desired foreign currency whilst already, right from the start, these industries will call for considerable import savings.

"An ideal solution is naturally a simultaneous development of heavy industry together with the planning and establishment of medium to large sized industries which are based on natural renewable resources, i.e. industries that use raw material derived from agriculture, including animal husbandry, forestry and fisheries. Depending on the type of raw materials, most of these industries can be developed economically in medium-sized plants. However, for certain commodities, e.g. pulp and paper and fish canning, the plants have to be of a larger size. Some commodities can also be processed economically in small units.

"Especially when propagating the use of smaller units, policy makers sometimes tend to belittle these types of industries. They are less spectacular, and it is difficult to include them in statistics which aim at 'counting chimneys'. However, the greater number of newly developing countries are basically dependent on agriculture, animal husbandry, forestry and fisheries, and the majority of their people still live in a rural environment.

"Industrial development based on the natural renewable resources means in effect *rural* development, which is a must, not only in order to stop an unhealthy, often unethical and uneconomic rate of urbanization, but above all to improve the social conditions and economic productivity of the majority of the population. In such development projects and acceleration of the national rate of economic growth and a rise in the volume of the gross production will soon be shown.

"The most usual path to progress is obviously first of all to increase food production. As efficiency in food production increases, the proportion of the population needed to produce food will decrease. Establishment of selected food and food products processing pilot plants encourage in turn the farmer to grow more and better food. In addition to selling his crops fresh at the local market, the farmer is assured of having an outlet for his produce at fixed prices and even more profitable one (guided by the industrial plant which can only operate economically by using raw material of the quality required by the technological process that is applied) he can supply first quality produce.

"Food and food products industries not only encourage local farmers to grow more and better quality produce, they also reduce losses and often a combination of waste and by-products can be the beginning of another industry, e.g. the processing of animal fodder. If carefully managed, the prices of food and food products, through semi or full processing, improved storage and packaging, will decrease, and preserved food and new food products will then also become available for urban areas.

"Naturally, in order to reach the point when the percentage of the household budget for food and food products will be as low as it is at present in advanced countries, the establishment of food industries is only the first step. In addition, especially for larger villages and towns, a type of supermarket facility or similar marketing units will be needed, adjusted to the local environment, demand and food habits. New food and food products based on local taste, will gradually develop. These new food products will, in the beginning, create a problem for the consumer, who needs to be educated as to how to buy selectively with a view to a balanced nutritious diet for the family. As a result, the establishment of food and food products industries will raise the income of the primary producers, and consequently open up possibilities for manufactured commodities which farmers can afford. Undoubtedly, at a later stage such manufacturing plants catering for a local demand will also be established in the rural areas.

"In turn, there will be less pressure on agricultural activities, and labour will become available to operate these industries. In short, as there is not much chance of bringing about any significant improvement unless the productivity of rural workers is increased, a rather complex series of actions will be required. Rural industrialization means an even spread of industries through the countryside, based on local resources, both in terms of human skills as well as raw materials and local needs.

"Rural industrialization needs to be one of the major economic and social objectives in development planning, and the aim should be to convert the agricultural communities to commercial farming, and then to well-balanced and diversified agro-industrial activities, where money is made from selected industries linked with agriculture in its broadest sense (including animal husbandry, forestry and fisheries). Such an industrial system will considerably reduce losses all along the line from the primary producer through the processing plant, storage and market facilities, to the consumer. Losses can only be controlled at these levels once such industrial systems with storage and market facilities are established, whilst such a chain is the only one that permits an economic use of by-products and waste-products.

"There is one other aspect that must be taken into consideration and that is the effect that industrial pilot units, storage and market facilities have on the education at the local technical level. Such units, through demonstration and training, raise the technical, technological, and managerial level and are, therefore, in fact, the best investment in human skills, which is the most productive form of employment on which larger future industrial projects can be built.

"In conclusion, agricultural and industrial development in many instances are only different terminologies that apply to the same processes and activities to promote economic growth throughout the widest possible segment of the population. In many spheres the two are fundamentally identical and, therefore, have to supplement and reinforce each other."

Plan of Operation

A Special Project aimed at Increasing use of Renewable Energy Sources in India

Special Project Allocation	
Duration :	25 years (concentrated effort: 10 years)
Executing Agency:	Special Project Agency (Volunteer Action Group)
Cooperating Government Agency:	Department of Science and Technology, Ministry of Energy

This project is to promote increased use of renewable energy sources such as sunlight, wind, biogas, fuelwood and charcoal-geothermal and hydro-electric power, animal draught power and other sources which are renewable and perpetual as against fossil fuels, such as petroleum and coal which, once utilized, cannot be replenished. The Special Projects Agency will implement, on an emergency basis, its Plan of Operation which is pledged by the Volunteer Action Group, and for which the Government of India is willing to take appropriate steps, including stimulation of large-scale action through grants and other means.

Background and Justification

India's endowment with energy systems is by no means large. There are also obvious limitations to the extent to which savings in energy consumption could be achieved through conservation. It is apparent, therefore, that if the country's growing demand of energy is to be met, new and *renewable* sources of energy must be developed. The new energy technologies are particularly suited for the introduction of decentralized or small-scale energy supply systems which fit in with India's rural agricultural economy. Remote communities which cannot be reached by electricity grids except at prohibitive costs, or which do not have easy access to conventional commercial fuels, could through renewable energy technologies be provided with energy sources for domestic use as well as for rural industries and agriculture. In general, these technologies have the additional merit of having less adverse impact on the environment than the conventional fossil fuel systems. Even though the immediate contribution of the new energy technologies to the energy supply may be limited and many of them are not economically competitive at the moment, there is no doubt that they will play a predominant role in the future.

Purpose

The purpose of the project is to promote increasingly efficient use of new and renewable sources of energy, including stimulation of set targets for India. The broad approach is as follows:

(a) To implement large-scale projects, such as those of energy, forestry and biogas, where technology development has already reached a stage which permits field application.

(b) To carry out field testing and demonstration on a country-wide basis of technologies in the field of renewable energy.

(c) To assist in carrying out research, design and development of devices such as :

- solar cookers
- solar crop driers
- solar water heaters
- solar pumps
- solar refrigerators/coolers
- solar space heaters
- solar energy storage devices.

(d) To promote the use of windmills, especially for water lifting, both for irrigation and drinking water, in particular in areas of suitable wind velocities.

(e) To promote hydro-power and more specially mini hydro-plants.

Description of Programmes

Below follows a description of some of the programmes of GOI in the field of renewable energy which must be supported by this Special Project.

Energy Forestry

In the emerging energy situation, as stated earlier, the role of non-commercial fuels acquired a new significance. Rather than proceed on the assumption that substitution of non-commercial by commercial fuels is inescapable, a systematic effort has now to be made to sustain the supplies of such fuel and to improve the efficiency with which they are utilized. Firewood, as is well-known, is the most important among the traditional of fuels, accounting for almost two-thirds of the total energy contribution from non-commercial sources. It is becoming increasingly scarce and in many parts of the country, it has already ceased to be non-commercial in the strict sense. The ecological damage caused by extensive denudation of forest land for fuel and other requirements is too well-known to need any elaboration here. There is, no doubt, the problem of pressure on land and the competition for land use for agricultural, industrial and housing activities. Recent experience, however, shows, that even in the thickly populated areas of the country, sufficient land is available along roads, railway lines and canals and in the degraded forests and community lands for developing wood lots to provide fuel to the local communities. Energy forestry would have to be regarded as an important component of our energy strategy, as the rural communities will continue to depend heavily on firewood for several decades to come. A large programme of fuel and farm forestry is, therefore, being taken up in the Sixth Plan, the target being set at 13 lakh hectares of plantation. The details have been set out in the section on "Forestry" in the chapter on Agriculture.

Biogas

The technology for conversion of animal wastes into biogas is well established and a fairly large number of biogas plants are already in operation. While the pioneering work in this field has already been carried out by the Khadi and Village Industries Commission, a number of agencies are currently involved in a national coordinated programme with the research effort aimed at raising tbe efficiency of gas generation and bringing down the capital cost of the biogas units.

Some new designs have already become available, for instance drumless biogas plants, which do not require steel for construction and are built of bricks, are proving popular in Uttar Pradesh. Ferro-concrete is another material which holds out possibilities of displacing steel and reducing cost in the construction of biogas plants. Some innovative solutions to the problem of poor gas generation under conditions of low ambient temperature also appear to be in sight. Further research efforts will be directed at cost reduction through improvement in design and use of materials, as also fermentation and the use of crop residues and water hyacinth in biogas production. Problems relating to the use of biogas for providing motive power and electricity will also be taken up.

The biogas programme, as is well-known, not only provides much needed energy to the rural families reducing the pressure on firewood supplies, but yields valuable organic manure, improves the environmental health in rural areas and reduces the drudgery of women. Having regard to the large potential which this programme offers of developing local sources of energy supply, an ambitious programme of installation of biogas plants is being launched. About one million family size plants and 100 community plants are proposed to be set up during the Sixth Plan. While financial provision will be made in the Plan for paying appropriate subsidies to those who install the biogas units, institutional funds will have to be harnessed for the prospective owners to meet the initial capital cost. It will also be necessary to strengthen the machinery for providing technical guidance, assistance in installation and post-commissioning service in the rural areas, and to initiate a programme for dissemination of information relating to biogas and its applications.

Solar Energy

For a tropical country like India, solar energy could be an abundant source of renewable energy. The major problem in harnessing solar energy is the fact that it is diffuse and variable. While solar energy can be utilized successfully for low heat applications in which large variations in energy output are not critical, the technologies for concentrating solar energy for high heat applications, or where energy storage is needed in order to provide a steady output, are still rather expensive. During the last few years, a broad-based coordinated solar energy research programme has emerged with an integrated approach.

An integrated approach for the use of biogas plants, firewood plantations and solar cookers is highly desirable. A comparison has been attempted in the integrated use of these sources of energy which yielded the results described below.

Comparison

The capital expenditure per person to meet the fuel need with different programmes as against cost of firewood per person are as follows:

Yearly cost of fuel required per person : Rs. 75.00
(300 kg wood at Rs. 0.25 per kg)

Programme	*Capial expenditure per person*	*Remarks*
Biogas Plant	Rs. 500	Valuable manure residue is a major advantage.
Firewood Plantation	Rs. 50	Area required for 30 persons is one ha. Five to seven years are required to get yield.
Solar Cooker	Rs. 50	Return on savings of fuel are immediate.

Both programmes—biogas plants and plantations—are vital from different considerations. The comparison, however, brings out clearly that a national programme of solar cookers is not only economically viable but essential to meet the energy needs of cooking in villages, the reasons being:

(i) the capital cost per person is low and benefits become available immediately;
(ii) a national programme of solar cookers, starting with areas of low rainfall, would help to check the increasing demand for firewood; this would indirectly help the plantation programme;
(iii) wood from plantations could be diverted as raw material to more profitable uses; and
(iv) as a long-term measure, use of solar cookers on a wide scale would be preferable ecologically to burning wood for cooking.

11 A Plea for Quality

What is Quality: from Latin, *qualitas;* French, *qualité;* how constituted: that which constitutes proper or essential being, character, nature, kind. Specifically excellence of character, natural superiority in kind. Also an acquired trait, accomplishment, requisition.

Quantity: Latin, Italian *quanto*; How much: An amount of portion either a measurable or numerable amount, or loosely any amount capable of increase or decrease.

We are fond of displaying our quantity, to show that we are a big nation, a nation of 683 million people. Our plans are big, we talk big. We have produced 130 million tonnes of foodgrains, we produce 23 million tonnes of steel including steel ingots, saleable steel and steel casting *(India Year Book 1979)*. We have built so many dams each having so many million cubic metres of water. We have started so many community development blocks, a large and impressive number.

We know that number or quantity is important, but we should not forget *Quality*. Let us produce quality children, give them good quality of education, so that each one will be worth much more. In terms of food production, let us not only talk of producing rice and wheat, but let us produce more grapes, apples, vegetables, pulses, beans, milk and milk products, meat and meat products, etc. This is quality food.

We talk of 105 Universities plus 10 Institutes deemed to be universities, several thousands of colleges and several hundred thousand of schools. It is good to spread education, learning, knowledge. But knowledge is power only when it helps in some production and furthers right action. To know how to grow melons is of no value unless you grow them, or teach others by growing them. Let us have practical schools for agriculture, technology, for trades and professions. Often agricultural schools and colleges have not enough or even any land where students can grow crops. Imagine, the diploma or degree which is supposed to indicate that you know how to do a particular work which becomes an end in itself and not the knowledge or experience that you get from this education.

The quality of men and material is important. We should build houses which last, roads which will not be easily washed away or dams which will not be silted up and become useless. We build dams, large canals and irrigation network without fully ensuring that every bit of the water stored or canalized is fully utilized. This is like buying a lot of bread when you are hungry, consuming what you need immediately and wasting the rest. This is happening with the irrigation canals in many places.

Even on the food front, there is information that if only half the losses during harvesting and storage of foodgrains are eliminated, we will have enough food to both consume and export. We produce the food but we do not have it (because we waste it).

By concentrating on the quality, the implication is that each operation has to be given adequate emphasis. It is a matter of detail that one goes into that makes for quality. A nation or people are respected much more if there is quality. We have resources and brains. Let us muster the will to produce more of good quality things. At this stage of national emergency we need more and more quality and our past achievements show that we can do it.

12 The Organisation - HIMCON

Himalaya Consortium for Himalaya Conservation

HIMCON is a non-profit voluntary organisation which was initiated by its founding Chairman Dr. M.L. Dewan more than 15 years ago under the banner of TFL-HIMCON. It was later registred in New Delhi as an independent charitable trust called HIMCON initiated by Mrs. Sudarshana Dewan in the year 1999. The Dewan family started its Himalayan networking from their home at Shivranjini later also called as Himalaya Sanrakshan Ashram at Ranichauri in the Tehri in the Tehri Garhwal district of Uttaranchal (now Uttarakhand) and developed it as its field headquarter for exchange, development and implementation of new ideas towards improving the quality of life of the Himalayan people and its environment.

The Region

HIMCON has been working in the Uttarakhand region of the central Himalayas and has focused its programmes and activities in the Tehri Garhwal district although various activities have been taken up in the other districts also in collaboration with like minded partner organisations and village communities.

Sharing international frontiers with Nepal and Tibet (China) Uttarakhand is strategically located and has two administrative divisions (Garhwal and Kumaun) and thirteen districts. Eastern part of the state (Kumaun) remained under direct British rule while a large portion of the western part (Tehri-Garhwal) was a protectorate state ruled by the native provinces. Present Tehri-Garhwal district represents a part of the latter and is surrounded by Pauri, Uttarkashi, Rudraprayag and Dehradun districts. Though drained by Bhagirathi and Yamuna rivers the district faces severe crisis of water. Despite hosting mighty Tehri Dam a large population of the district does not have access to eletricity. Literacy and employment rates in the district are very low.

The beauty of the Himalaya, the abode of eternal snow, has attracted explorers, adventurers, peace and nature lovers since times immemorial. The scenic beauty of the terrain, is biodiversity, and colourful people, their distinct cultural and ethnic identity, has been highlighted both in literature and mythology. Himalaya represents the northern mountainous frontier of our nation and it stands testimony to the creative beauty of Mother Nature that gifted us with this majestic landscape out of the oceanic debris of Tethys Sea, in recent geological past.

Though young, Himalaya provided basic ingredients for the growth of rich and diverse cultures in its lap. Thriving in this terrain for quite long these created a distinct identity for their community that incorporates basic precepts of coexistence with nature with thorough understanding of the principles of ecological conservation. Economy, as also the religio-cultural life of these people traditionally revolved round the surrounding pristine forest and pastoralism; agricultural being confined mainly to the river valleys.

The ongoing geo-tectonic movement together with isostatic adjustments along this young mountain chain are a cause of concern as these make terrain prone to natural disasters and mass wastage. Anthropogenic interventions further aggravate the pace of these degradational processes.

The inhabitants, through experience and experimentation, learnt to overcome the handicap imposed upon them by the terrain and evolved the art of locating and exploiting subsurface waters and the habitations were built around these. The traditional acumen of the people is unparalleled even today. Exploiting the bounty of

Mother Nature the people embarked upon an altogether new line of economic activity that was non-land based. Economic strength of these cultures is reflected in majestic structures that were supported by its surplus produce and these owned by even the common people.

The cultures of the Himalayan terrain set forth foundations of a strong system of resource management and evolved genetic craftsmanship for developing whole range of farm and animal species so well suited to the agro-climatic conditions and physiography of the region.

The self-content inhabitants of this land were happily living in their heavens until the gates of their land were flung open and armed with the disparity in purchasing powers the new settlers acquired rights over the resources of this virgin land. A welfare state would have interfered at this stage to conserve the resources and rights of the natives, but the interests of many in power forced the state to stand testimony to this blatant exploitation. The natural resources of this terrain were thus plundered in a never before pace and the repercussions of these are reflected in enhanced frequency of natural disasters and high rates of mass wastage that the terrain experiences at present. Lately, the inhabitants of the downstream plains have also started to experiences the impact of these unplanned interventions in the mountains. With increasing contacts with the world outside the social fabric of the society loosened and this weakened the traditional and time-tested systems of resource management.

Once surplus producing economy of the region was thus transformed into the one dependent largely upon outside remittances. As a reaction to the changes taking place and in a bid to strengthen their economy and also to improve their quality of life the ablebodied personnel of the region resorted to out-migrate in search of jobs, so as to create surplus to be pumped into the cash starved economy of the region. Though well intentioned, this option further destabilised the economy and the region faced shortage of farming hands together with resource mismanagement, depleting fertility, increasing erosion, depleting water availability, increasing drudgery of the females. The reality is that the region faces severe dearth of working hands, brains, money, and youth. Serious and concentrated efforts are needed to change this trend that seems to have crept deep into the lives of the masses at grassroots level.

Out Focus

The HIMCON team is dedicated to the Protection and Conservation of Himalayas and focuses on:

- Improvement of the quality of life of the masses of this region through small initiatives originating from within the society and mostly supported by it.
- Greening the Himalayan terrain.
- Organic farming and Horticulture Development.
- Propagation of medicinal plants and herbs.
- Water harvesting and conservation.
- Providing safe drinking water to the village communities.
- Improving the nutritional value of the food available to the masses.
- Land slide mitigation.
- Value based education.
- Vocational Training.
- Traditional health care and nature care.
- Motivating youth and women to participate in various development activities.

Moved by the drudgery of the women HIMCON also organised a special drive aimed at improving their conditions at the grassroots level. Designated as "Mahila Shakti" project, under the leadership of CHIRAG (Central Himalayan Rural Action Group) this programme sought to empower the women, who is the pivot of all farm and household activities and is over-burdened with work.

Table 12.1. Details of the plantation undertaken by HIMCON in the developmental activities

	Year							
	1992		1993		1994		1995	
	Multipurpose trees	Fruit trees	Multipurpose trees	Fruit trees	Multipurpose trees	Fruit trees	Multipurpose trees	Fruit trees
GIC Ranichauri	8,096	2,925	2,800	2,460	–	–	2,500	110
UJJs Khari	1,020	4,119	–	–	–	–	–	320
GIC	–	25	5,000	–	–	5,450		
MMD Sabli	–	–	5,050	327	–	–	–	–
PJKS Lambgaom	–	–	–	–	–	–	–	30
MNS Chamba	–	–	5,000	–	–	–	–	150
GIC Gaja	–	–	450	200	–	–	–	–
MMD Dargi	–	–	–	550	–	–	–	–
MMD Jagdhar	1,000	500	–	300	2,000	300	500	400
Total	10,106	7,569	18,300	3,837	2,000	5,750	3,000	1,010

Initial phase of HIMCON was a period of learning and important lessons learnt in this phase while working with the masses have been carved out boldly in the philosophy of the organisation. Appreciative of the beauty of traditional system of resource management together with that of the social fabric of the people habitating the terrain HIMCON formulated a blueprint of strengthening the traditional systems through limited technological inputs and of course, people's participation. With time a dedicated programme launched by the chairman of HIMCON was formed with thorough understanding of the ground realities at the grassroots level and the success of various projects of the Organisation is owed to the team spirit.

These early experiences and lessons prepared ground for taking up developmental endeavours in a planned manner. People's participation was the starting point and even after years of experience at grassroots level it still remains the keyword for the success of all our programmes. People's participation has been the central theme of all the initiatives despite working in close co-operation of a large number of experts of diverse fields and employing state of art technologies in our quest to impove the quality of life of the masses. Though proud to be utilising the latest of GIS and Remote Sensing technologies for better planning and management of the resources, we have no hesitation in accepting the truth that our strength lies in the support we get from the masses for our programme's implementation.

HIMCON has put in more than a decade of community service for the cause of the Himalayan region and has also highlighted the problems of the masses before the national mainstream through the audio-visual and print media.

Our Programmes and Activities

1. Awareness Drive.
2. Community Groups.
3. Mahila Mangal Dal (MMD).

Table 12.2: Details of MMDs and SSSs facilitated in the region by HIMCON

Sl. No.	*Village*	*Membership of MMD*	*Revolving fund of the SSS (in Rs.)*
1.	Paturi	67	55,000
2.	Lawadhar	23	28,000
3.	Copriyali	28	35,000
4.	Bidkot	22	39,000
5.	Daul	07	26,000
6.	Bhatusain	38	12,000
7.	Palash	38	15000
8.	Faigul	45	35,000
9.	Vaily and Ramolgaon	26	29,000
10.	Vajawar	18	23,000
11.	Kot Maniyar	105	1,23,030
12.	Muriya Gaon	49	57,000
13.	Dadur	46	46,483
14.	Pali	28	15,000
15.	Gajana	16	12,000
16.	Koti Gad	52	900
17.	Dharkot	71	1100
18.	Kirgani	45	900
19.	Dang Talla	23	460
20.	Chhati	45	16,000
21.	Kothi	26	23,000
	Total	827	5,92,873

4. Swayam Sahayata Samuha (SSS)/Self Help Group.
5. Human Resource Development : Trining Programmes.
6. Human Resource Development.
7. Vocational Training Centre, Renichauri.
8. Aarogya: The green enterprise.
9. School Environment Education Programme.
10. Environment Education Programme.
11. Value Education Center, Sabli.
12. Water Management.
13. Silviculture.
14. Horticultural initatives.
15. Community plantations.
16. Offason Vegetable Cultivation.
17. Health and Sanitation.
18. Landslide Management.
19. Detailed survey of the entire area falling in the reservoir rim of Tehri dam.
20. Animal Husbandry.
21. Alternate energy options.
22. Technological Adaptation.

Table 12.3 : Details of the Sewing Centres in the region

Sl. No.	*Village*	*Number of Sewing Centres*
1.	Paturi	01
2.	Birkot	01
3.	Bhatusain	01
4.	Falgul	01
5.	Vaily and Ramolgaon	01
6.	Vajawar	01
7.	Kot Maniyar	03
8.	Muriya Gaon	01
9.	Dadur	01
10.	Chhati	01
11.	Kothi	01
	Total	13

Table 12.4 : Vocational training programmes organised by HIMCON

Sl. No.	Subject	Number of trainees	Beneficiary farmers
1.	Horticulture	12	625
2.	Vegetable cultivation	7	230
3.	Medicinal plants cultivation and processing	6	398
4.	Food processing	1	16
	Total	26	1,269

Table 12.5 : Details of interventions undertaken in the schools

Sl. No.	School	Number of Plants	Provision for water storage	Boundary wall (in metres)
1.	GIC Gaja	300	1	300
2.	GIC Nakot	470	1	–
3.	GIC Nagni	460	1	300
4.	GIC Pokhri	161	1	–
5.	GIC Dhaniakot	300	1	–
6.	GIC Nagarasudhar	150	–	–
	Total	1,841	5	600

Table 12.6 : Details of the water related facilities created in the region by HIMCON

Sl. No.	Village	Ferrocement	Spring Rainwater Harvesting	Chals water Tank	Canal (in metres)	Check Dam
1.	Patudi	07	03	02	–	03
2.	Lawa Dhar	03	01	01	–	03
3.	Copriyali	08	–	–	–	–
4.	Bidkot	06	–	01	–	–
5.	Doul	03	–	–	–	–
6.	Bhatusain	02	01	–	–	–
7.	Palash	07	–	–	–	–
8.	Sabli	03	03	–	–	–
9.	Faugul	13	05	01	300	–
10.	Vaily & Ramolgaon	04	01	–	–	–
11.	Vatan Khet	01	–	–	–	–
12.	Vajawar	03	01	–	–	–
13.	Kot Maniyar	41	03	–	–	–

(Contd.)

(Contd.)

14.	Muriya Gaon	14	01	·	·	02
15.	Dadur	13	–	–	–	–
16.	Pali	03	–	–	200	–
17.	Gajana	02	01	–	–	–
18.	Koti Gad	–	–	–	180	–
19.	Dharkot	07	–	–	–	–
20.	Kirgni	04	–	–	–	–
21.	Dang talla	12	–	–	–	–
22.	Chhati	–	01	–	150	–
23.	Kothi	04	02	01	150	–
	Total	185	23	06	1,480	24

Table 12.7: Horticultural plants in the area by HIMCON

Sl. No.	*Village*	*Fruit plants*
1.	Paturi	2000
2.	Lawadhar	500
3.	Copriyali	1000
4.	Birkot	1000
5.	Daul	1000
6.	Bhatusain	500
7.	Palash	1000
8.	Sabli	5000
9.	Faigul	500
10.	Vaily and Ramolgaon	250
11.	Vatankhet	75
12.	Vajawar	250
13.	Kot Maniyar	3164
14.	Muriya Gaon	–
15.	Dadur	200
16.	Pali	1500
17.	Gajana	300
18.	Koti Gad	436
19.	Dharkot	–
20.	Kirgani	–
21.	Chhati	400
22.	Kothi	
	Total	19,675

Table 12.8: Details of Saheed Subab Sing Sajwan community plantation

Sl. No.	*Planting season*	*Plant species*	*Number of plants*
1.	Winter	Plum	360
2.	Winter	Peach	200
3.	Winter	Pomegranate	100
4.	Winter	Pear	300
5.	Winter	Pecan	300
6.	Winter	Almond	210
7.	Winter	Popalar	1,000
8.	Winter	Agave (for fencing)	10,000
9.	Monsoon	Guava	200
10.	Monsoon	Orange	300
11.	Monsoon	Citrus	100
12.	Monsoon	Surai	100
	Total		13,170

Table 12.9: Details of the Pali plantation

Pomegranate	200
Pear	200
Apricot	200
Peach	150
Plum	100
Almond	75
Walnut	65
Kiwi	95
Total	1,085

Table 12.10: Details of the production of snow peas in the previous season

Sl. No.	*Village*	*Beneficiary farmers*	*Production (in kg)*
1.	Chorgarh	6	68.9
2.	Saur Kot	1	4.1
3.	Tikhon	1	2.3
4.	Thangdhar	12	170.9
5.	Saur	4	39.0
6.	Aniargaddi	3	33.2
7.	Pataldevi	2	39.0
8.	Pali	30	396.0
9.	Rauslikhal	10	144.9
10.	Koti Gad	12	216.0
11.	Dharkot	7	96.0
12.	Kirgani	11	126.0
13.	Dang Talla	15	33.0
14.	Sangaon	8	148.6
15.	Badalsain	6	79.9
16.	Kureth	18	316.6
	Total	146	1,914.4

Table 12.11: Health camps arranged in the region by HIMCON

Sl. No.	*Nature of camp*	*Number of camps*
1.	General check up camps	20
2.	Eye camps	7 (200 Corrective surgeries undertaken)

Table 12.12: Details of the latrines constructed in the region

Sl.No.	*Village*	*Toilet*
1.	Paturi	45
2.	Lawadhar	15
3.	Copriyali	07
4.	Birkot	05
5.	Daul	03
6.	Bhatusain	04
	Total	79

Table 12.13 : List of the direct beneficiaries to Aarogya Dugdhashala

Sl.No.	*Beneficiary Name*	*Village*
1.	Smt Kawara Devi	Jardhar Gaon
2.	Shri Ratan Singh Negi	Bidon
3.	Shri Karam Singh Bisht	Bidon
4.	Smt Bachni Devi	Chopriyali
5.	Shri Purnanand Dobhal	Sabli
6.	Shri Harish Dobhal	Sabli
7.	Shri Surveer Singh	Basoie
8.	Shri Gobind Pd. Dabral	Matali
9.	Shri Kunwar Singh	Patudi
10.	Shri Rajendar Singh	Jardhar Gaon
11.	Shri Lokendar Saklani	Deiori
12.	Shri Mahaveer Singh	Khankar

Table 12.14 : Details of plantation at Kanatal at Kanatal Farm in 2004-05

Sl.No.	*Plant*	*Species*	*Number*
1.	Walnut	*Franquette*	*100*
2.		*Vina*	50
3.		*Hartley*	100
4.		*Chandlar*	50
5.	Apple	*Vance Delicious*	100
6.		*Red Chief*	50
7.		*Oregon spur*	100
8.		*Silver spur*	100
9.		*Top Red*	50
10.		*Red Delicious*	50
11.		*Tydoman Warcesster*	50
12.		*Red Gold*	50
13.		*William Spur*	50
14.		*Royal Gala*	300
	Total		1,200

Table 12.15 : Details of plantation at Kanatal Farm in 2004-05

Sl. No.	*Plant*	*Species*	*Number*
1.	Apple	Oregon spur	30
2.		Top Red	10
3.		Royal Delicious	80
4.		William Spur	80
	Total		200

Table 12.16 : Details of the Horticultural plants introduced around Kanatal in 2004-05

Sl. No.	*Village*	*Farmers*	*Details of the plants distributed*			
			Apple	*Walnut*	*Almond*	*Total*
1.	Chopariyal Gaon	27	3,198	–	–	3,198
2.	Kainthogi	10	3,200	500	200	3,900
3.	Chunnipani	04	1,200	–	–	1,200
4.	Jari Pani	27	3,502	100	–	3,602
5.	Patal Devi	08	1,615	–	–	1,615
6.	Thang Dhar	21	3,200	–	–	3,200
7.	Chari Palda	07	3,200	500	300	4,000
8.	Semal Patal	19	1,600	600	500	2,700
9.	Nair	19	4,000	700	–	4,700
10.	Ransuli Khal	17	1,588	–	–	1,588
11.	San Gaon	09	2,312	200	–	2,512
12.	Kanatal	14	1,500	–	–	1,500
13.	Dev Darshani	07	1,100	–	–	1,100
14.	Kadukhal	07	1,400	–	–	1,400
15.	Thuner	07	2,600	–	–	2,600
16.	Bhamorikhal	08	800	200	–	1,000
17.	Bharodhunga	17	3,135	200	–	3,335
	Grand Total	228	39, 150	3,000	1,000	43,150
Land area under horticulture			98 ha	15ha	2.5 ha	115.5 ha

ANNEX-1

PARTNER ORGANIZATIONS

Trees for life, New Delhi
Sudarshana Welfare Trust, New Delhi
Himalaya Sewa Sangh, New Delhi
Mahila Haat, New Delhi
Society for Promotion of Wasteland Development, New Delhi
CHIRAG, Nainital
INHERE, Masi, Almora
Shaikshanik Grammonnati, Pithoragarh
Swati Gramodyog, Pithoragarh
Uttarakhand Jan Jagriti Sansthan, Tehri-Garhwal
Parwatiya Jan Kalyan Samiti, Tehri-Garhwal
G.B. Pant University of Agriculture and Technology, Pantnagar
Hill Campus of the G.B Pant University of Agriculture and Technology, Ranichauri
Kedia Exports Flex Foods Limited
Vijay Garg
Chait Ram Dhani, Himachal Pradesh
Venu Eye Research Institute, New Delhi
Himalayee Paryavaran Shiksha Sansthan, Uttarkashi
Laxmi Ashram (Kasturba Mahila Utthan Mandal), Kausani-Almorah

HIMCON acknowledges with thanks to:

- Trees for Life
- Tata Tea Ltd., Bombay
- Foundation for International Training (FIT), Toronto, Canada
- Food and Agriculture Organisation of the United Nations
- Sir Dorabji Tata Trust, Mumbai
- The Dewan Foundation, London

- Indo-Global Social Service Society, New Delhi
- The Ford Foundation, New Delhi
- Department of Science and Technology, Government of India, New Delhi
- Ministry of Non-Conventional Energy Sources, Goverment of India, New Delhi
- National Horticulture Board, Government of India, New Delhi
- Rotary International
- Rotary Club of Delhi South Metropolitan
- Neri

And to all our friends and co-workers for extending their valuable support and co-operation in impelmenting our mission to protect and conserve the Himalayas and serving the Himalayan communities.

ANNEX-2

HIMALAYA CONSORTIUM FOR HIMALAYA CONSERVATION

HIMCON

Head Office	Dr. M.L. Dewan, N-44, Panchashilla Park, New Delhi-110 017 (India)
Tele Fax	+91-11-26496916
Cell Phone	9810149503
E-mail	mldewan79@hotmail.com
Website	www.himcon.net
Field Office	HIMCON Milan Kendra Sabli, Tehri Garhwal-249 199 (India)
Phone	+91-1376-252070
Cell Phone	9719583170
E-mail	rakeshbahuguna123@rediffmail.com

Legal stature	Charitable trust registered at Delhi
Registration number	388/7193/2.09.1999
FCRA number	231660090
35 AC number]	
Trustees	Dr. M.L. Dewan, Manoj Pande, Radha Behn, Neerja Chaudhary, Ravi Gulati, Dr. Piyoosh Rautela, Rakesh Bahuguna
Settlor	Mrs. Sudershna Dewan

Himalaya consortium for Himalaya conservation (HIMCON) is a non-profit organisation committed to the cause of Himalayan environment and the welfare of its inhabitants. We value the rich cultural traditions of this terrain and seek to promote traditional systems of resource managments. We welcome contributions of all sorts that would strengthen our pursuit of serving the Himalaya.

HIMCON Activities

(i) Vegetable cultivation
(ii) Training programme
(iii) Horticulture Plantation
(iv) Mahila Mangal Dal (MMD)
(v) Rain water harvesting
(vi) Swayam Sahayata Samuha (SSS)
(vii) Training and Exposure Programme
(viii) Horticulture Development Programme
(ix) Vegetable Seed and Seedlings Distribution
(x) Soil Conservation and Water Management
(xi) Water Conservation
(xii) Recharge Augmentation
(xiii) Irrigation Facility Creation.

APPENDIX-I

State of Himalayas 1986 : A Call for Action

Summary and Recommendations

1. The Himalayan Eco-System is one of the most important and most threatened of life support systems on earth. In the shadow of the Himalayas, live more than 270 million people in India, some of them poorest in the world. The rivers which arise in the Himalayas and flow down into the wide Indo-Gangetic plain support the essential agriculture which sustains these people. The population pressure on land and its mounting demand for fuel and fodder has already denuded areas of the Himalayan Ranges – the lifeline of India. Himalayan land resource degradation has been a matter of considerable concern in recent years. The Himalayas, our sacred mountains, are in very poor state of health due to deforestation, increasing pressures of human and animal population and their needs and greeds. The rate at which the forests are disappearing is much faster than the rate of afforestation.

2. The Government of India and the State Governments are making efforts to conserve the Himalayan environment, but are also realising the fact that they alone cannot undertake to carry out such a gigantic task.

3. Work by the Ministry of Environment and Forests, National Wasteland Development Board, Soil and Water Conservation Division of the Ministry of Agriculture, Department of Non-Conventional Energy Sources and other related government departments are still not able to carry out more than a fraction of

what is needed in the Himalayas. It is urgent that people's participation must be promoted for such a Himalayan Eco-System Programme. The objectives, principles behind the action plan and the framework for action are given in paras 7, 8 and 9.

4. In the call for action, six pronged action programme which in brief asks for increased tree plantation, decreased deforestation, improving springs and water harvesting, promotion of renewable sources of energy, integrated programme for improving crops, livestock, agro-forestry, grassland, biofencing, etc. and intensification of poverty alleviation programmes, etc. The strategy also includes identification of critical catchments and their repair, also an urgent call to the people and to the government for a concerted action.

5. The proposals for accelerated action identify:

(i) The problem
(ii) The success stories
(iii) Proposals for accelerated action
(iv) Recommendation for Five Year Programme (1987-91) including investment needs.

6. The 'State of Himalayas 1986 - A Call for Action' has been prepared by the authors as Member (Soil Conservation), National Land Use and Conservation Board, on invitation of the Board. The emphasis is on people's participation, first report of which was published in 1985. This followed with several programmes, eco-development camps, nurseries, plantation, formation of local conservation and forest societies in the Himalayas.

7. *Principle behind the action plan* - First and foremost, it has to be accepted by all concerned that the best and the most practical route for conserving natural resources (soil, water, vegetation, etc.) is through their wise and rational utilisation. The local communities place a value on the resources; it is precisely because of this that they are utilising them and are benefited by the use. What is important is not to alienate the local communities by dropping a

'legal curtain' between the people and the resources but to find ways and means of seeking people's participation for sustainable use of the resources. This plan, therefore, places the people at the centre of the stage and calls for management of the resources through them and by them.

8. The *objectives* of the call for action are:

(a) Through people's participation achieve:
   - the protection of the soil;
   - the promotion of employment of local communities who should be the partners in conserving and utilising the natural resources in a sustainable manner;
   - the acceleration of income and the improvement in the living standards of the people;
   - development and improvement of the availability of surface as well as ground water resources;
   - promotion of viable and sustainable farming systems which are integrated in nature and serve not only the comprehensive needs of the farming communities but are also environmentally sound.

(b) To ensure that floods are prevented thus contributing to the development of sustainable agriculture in tho low lands. The reduction of siltation of the rivers has manifold other benefits.

(c) Conservation of the uplands and the catchments will prolong the life of the major and minor reservoirs that are already constructed.

9. *The framwork for action*–In view of scarce resources, the focus should at least for the coming five year period be on the most severely degraded areas which are also most densely populated. A survey has been undertaken (using satellite imagery and aerial photography) which aims at identifying the specific sub-catchments, villages and the lands around them that need to be conserved and developed urgently. Many of these sub-catchments

are identified. NGOs and voluntary agencies are undertaking leadership in each catchments, tasks proposed and six pronged action plan outlined. After detailed action plan, the task is to act.

Twenty five *critical* watersheds are being selected for urgent and concerted action. Non-government and people's organisations will give greater attention to these critical watersheds which are identified as giving greater damage in flooding, erosion and siltation. Government can continue its plan but in whatever way it can help the local societies it will be extra effort. These twenty- five critical catchments are based on detailed soil surveys as well as use of remote sensing imagery and aerial photographs. These are the critical catchments which are defined as "where the quantitative studies have shown high values and needing urgent attention due to high erosion evidence by the following data":

(i) Sediment yield index weightage values use (potential soil detachment).
(ii) Delivery Ratio representing percentage of silt detached reaching the reservoir site (or potential reservoir site).
(iii) High slope percentage, shallow depth of soil, texture of soil, soil calcareousness, graveliness, soil cover, land surface/ characteristics, existing erosion conditions such as gully, ravine and landslide conditions, etc.

These 25 critical catchments are in the following states:

(i)	Arunachal Pradesh	1
(ii)	Assam	1
(iii)	Himachal Pradesh	6
(iv)	Jammu and Kashmir	3
(v)	Manipur	1
(vi)	Sikkim	2
(vii)	Tripura	1
(viii)	Uttar Pradesh (8 Hill Districts)	5
(ix)	West Bengal	1

(x)	Other North East States	2
(xi)	Others not yet clearly defined	3

Some of these critical catchments need more studies before delineating them exactly on the map and on the ground.

10. The "State of Himalayas 1986 – A Call for Action" is based on six pronged strategies as follows:

1. *Increase production and planting of trees*

    (a) Regeneration
    (b) Planting programmes of fuel, fodder, fruit purpose trees.
    (c) Bio-fencing/social fencing.

2. *Reduce consumption of woodfuel and cutting of forests*

    (a) Alternate energy (solar/biogas/wind, etc.).
    (b) Mini hydel
    (c) Smokeless *chulhas*
    (d) Measures for energy conservation.

3. *Special treatment of critically damaged areas*

    (a) Select critical micro/mini catchments
    (b) Treatment around villages/towns where damage is extensive
    (c) Landslides must be treated by special means (Institute planting juniper, other ever green, etc.).

4. *Revival/regeneration of springs and ponds/water harvested in small scale water dams+mini micro hydel+not large dams.*
5. *A call to people :*

    (a) Family planning is *Most Urgent.* Produce less children.
    (b) Take the challenge. Be self-reliant.

(c) Establish local conservation societies, for survival work cooperatively and not always depend on the government, start District Forest Associations, People's Nurseries, catchment development boards, etc.

(d) Organise into special sectors or special groups such as:
   - (i) Ex-servicemen
   - (ii) Students
   - (iii) NGOs/voluntary groups
   - (iv) Women
   - (v) Other youth.

These groups or societies should work cooperatively on nurseries, tree planting and other aspects of the strategy and action programme.

6. A call to the government

(a) Help those below poverty lines.

(b) Promote integrated farming.

(c) Utilise wealth of NGOs, voluntary organisations; military, para-military, border road organisations, ex-servicemen, etc.

(d) Promote people's and school nurseries.

(e) Assist in creating catchment development boards, district forest associations, village and *van* panchayats.

(f) Undertake training programmes in all aspects of Himalayan Eco-development.

(g) Speed up the construction of a Research, Development and Training Centre in the Himalayas like the one proposed Centre for Himalaya Conservation and Development.

(h) Decentralise decision making process.

(i) Have a Central Coordination Cell to stimulate adoption of a strategy for saving the Himalayas and to monitor and evaluate the on-going programmes in the Himalaya.

11. While considerable financial, moral and resource support is needed, groups like Dasholi Gram Swarajya Mandal, Chamoli, Garhwal (UP), Himalayan Eco-System Development, and Plantation, Utilisation and Reforestation Enterprise (PURE) combining retired UN, Forest, Army Officers and some young people and many other groups of people throughout the Himalayas are determined to save the Himalayas mainly through people's effort. The task is tremendous and at present it is being done on an *ad hoc* basis. A Central Coordination Cell or a Nodal Point whose objectives are mainly to stimulate, catalyse, prepare, monitor and evaluate projects is urgently needed. It is also proposed that a systematic and coordinated effort for eco-development in the Himalayas should include a computerized system which will indicate the inputs of the people's group, the government's as well as other organisations. A Himalayan Conservation, Research and Development and Training Centre as planned needs to be expedited. Himalayan Eco-Development Policy Board should review the progress at least once a year. The Himalayas are in critical stage of deterioration and only a concerted and integrated action by all concerned/interested is urgently needed. To save the Himalayas is to save Indo-Gangetic plain and in effect to save the country from an eco-disaster.

An Ode to Himalayas

HIMALAYAS - you are the abode of Gods.
Our northern great wall
You are our guide, our guard

Himalayas - you are home of snow
And of Glaciers
From where thousands of rivers flow

You are sacred, sublime
You are closer to Heaven
You are home of Badrinath, Kedarnath
Vaishnodevi, Amarnath where millions climb

Your peaks Pir Panjal, Nandadevi, Annapurna
Kanchanchunga, Kedarnath, Gangotri
Also Kailash, Mansarover all
Speak of your majesty, glory

You are Grand, Majestic, Serene, Inspiring
View from Almora Forest Guest House
In early morning sun or from Kausani and
Hundred other sites

You are giver of life, trees, springs
Flowers, medicinal plants and now
You are instead beginning of floods, misery
But mighty Himalayas - you are bleeding, crumbling
Too much pressure - too many demands put on you

People worship and adore you in
One breath and hurt you, injure you in next one.

Why this hypocrisy, this indifference?
Can this be reversed or is it too late?
It is late in my humble views
But not too late.

We all must work together young and old
Instead worshipping you with empty words
Take direct steps positive and bold

Plant more trees, nurse more trees
Cut less trees burn less trees
Promote integrated use of Himalayan resources
Revive the springs
Life blood of Himalayas
Remove poverty from your sacred land
A call to the people, the government
To join in these efforts

We pledge to save our sacred Himalayas
Sources of our Mother Ganga
And benefactor of our fertile now
Sick plains of Indo-Gangetic, Brahmaputra

We promise we all will work together
To bring you to the forested glory
Majesty, Greenness, Sacredness

M.L . Dewan

2 October, 1986

Index